VOLUME FIFTY SEVEN

ADVANCES IN GEOPHYSICS

VOLUME FIFTY SEVEN

ADVANCES IN GEOPHYSICS

Series Editor

LARS NIELSEN

Department of Geosciences and Natural Resource Management, University of Copenhagen, Copenhagen, Denmark

ELSEVIER

Amsterdam • Boston • Heidelberg • London
New York • Oxford • Paris • San Diego
San Francisco • Singapore • Sydney • Tokyo

Academic Press is an imprint of Elsevier

Academic Press is an imprint of Elsevier
50 Hampshire Street, 5th Floor, Cambridge, MA 02139, United States
525 B Street, Suite 1800, San Diego, CA 92101-4495, United States
125 London Wall, London EC2Y 5AS, United Kingdom
The Boulevard, Langford Lane, Kidlington, Oxford OX5 1GB, United Kingdom

First edition 2016

Notices
Knowledge and best practice in this field are constantly changing. As new research and experience broaden our understanding, changes in research methods, professional practices, or medical treatment may become necessary.

Practitioners and researchers must always rely on their own experience and knowledge in evaluating and using any information, methods, compounds, or experiments described herein. In using such information or methods they should be mindful of their own safety and the safety of others, including parties for whom they have a professional responsibility.

To the fullest extent of the law, neither the Publisher nor the authors, contributors, or editors, assume any liability for any injury and/or damage to persons or property as a matter of products liability, negligence or otherwise, or from any use or operation of any methods, products, instructions, or ideas contained in the material herein.

ISBN: 978-0-12-809533-1
ISSN: 0065-2687

Publisher: Zoe Kruze
Acquisition Editor: Alex White
Editorial Project Manager: Helene Kabes
Production Project Manager: Vignesh Tamil
Cover Designer: Greg Harris

Typeset by TNQ Books and Journals

CONTENTS

CONTRIBUTORS

M. Bastani
Geological Survey of Sweden, Uppsala, Sweden

S. Bazin
Norwegian Geotechnical Institute, Oslo, Norway

T. Dahlin
Lund University, Lund, Sweden

R. Frauenfelder
Norwegian Geotechnical Institute, Oslo, Norway

I. Gultepe
Environ. Canada, Toronto, Ontario, Canada; NOAA NSSL, Norman, OK, United States

C.M. Krawczyk
GFZ German Research Centre for Geosciences, Potsdam, Germany; Technical University of Berlin, Berlin, Germany

Y.M. Leroy
École Normale Supérieure, Paris, France; Total SA, CSTJF, Pau, France

B. Maillot
Université de Cergy-Pontoise, Cergy-Pontoise, France

A. Malehmir
Uppsala University, Uppsala, Sweden

H. Maurer
ETH Zurich, Zurich, Switzerland

K. Merz
Geo2X, Baar, Switzerland

R.D. Miller
Kansas Geological Survey, Lawrence, Kansas, United States

M. Pavolonis
NOAA/NESDIS, Madison, WI, United States

A.A. Pfaffhuber
Norwegian Geotechnical Institute, Perth, Australia

R. Rabin
NOAA CIMSS, University of Wisconsin, Madison, WI, United States

L.V. Socco
Politecnico di Torino, Torino, Italy

K. Suto
Terra Australis Geophysica Pty ltd, Brisbane, Australia

R. Ware
NCAR, CIRES, Boulder, CO, United States

PREFACE

Geophysics covers a broad range of research fields. This volume of *Advances in Geophysics* consists of three chapters that are topically very different.

Leroy and Maillot review and give the theoretical background behind "limit analysis." Moreover, they discuss the application of this approach in relation to understanding deformational mechanisms in connection with accretionary wedges. Furthermore, in addition to developing and providing the theoretical overview, Leroy and Maillot have focused on presenting their material in a pedagogic manner making their contribution well-suited for courses given at MSc- and PhD-level educational programs.

In a rich-illustrated and comprehensive contribution, Malehmir et al. review and discuss the state of the art in many aspects of high-resolution geophysical imaging of different areas and settings that are at risk of experiencing natural hazards now or in the future. They give examples from a number of case studies and present and discuss the applicability of several, very different geophysical methods. Data examples shown by Malehmir et al. include, but are not limited to, different types of seismics, ground-penetrating radar, and geoelectrical approaches, just to mention some of the many methodologies discussed in their chapter.

Gultepe et al. present and discuss light snow observations at cold temperatures and Arctic regions, and their effects on weather and climate simulations. This type of precipitation is important for predicting and understanding weather forecasting and climate change. They present measurements obtained by new and conventional instruments. Then, they state impact of missing light snow on contribution to total precipitation. Finally, they also discuss aspects related to accuracy of measurements and data quality control and highlight the importance of these aspects for cold weather and climate simulations.

Enjoy your reading of volume 57 of *Advances in Geophysics.*

Lars Nielsen

CHAPTER ONE

Mechanics of Accretionary Prisms and Fold-and-Thrust Belts Based on Limit Analysis

Y.M. Leroy[*,§,1], **B. Maillot**[¶]
[*]École Normale Supérieure, Paris, France
[§]Total SA, CSTJF, Pau, France
[¶]Université de Cergy-Pontoise, Cergy-Pontoise, France
[1]Corresponding author: E-mail: yves-marie.leroy@total.com

Contents

Advances in Geophysics, Volume 57
ISSN 0065-2687
http://dx.doi.org/10.1016/bs.agph.2016.09.002

Abstract

The first of the two objectives is to present the theory of limit analysis (LA) and its application for determining the collapse mechanism (CM) characterizing the complex, active fold within a fold-and-thrust belt or an accretionary wedge. The theory applies to fluid-saturated materials corresponding typically to upper crustal conditions. The potential of this approach is illustrated by comparing the various CMs, which could occur at the front of an accreting wedge, including friction, delamination, and compaction deformation mechanisms resulting in either straight or curvy ramps for a thrusting oriented either hinterland or foreland. Such hinterland thrusting combined with a change in depth of the active decollement would be typical of triangular zones. In the process, it is shown that LA reproduces and extends naturally the classical critical Coulomb wedge theory. The second objective is to document how the predictive potential of the kinematic approach of LA is combined sequentially with the geometrical constructions of fault-related folds. The resulting method is called sequential limit analysis and provides the means to determine, for example, the complete sequence of thousands of thrusts, each thrust being captured from its onset to its arrest because of the onset of a mechanically more favorable event.

1. INTRODUCTION

The prime objective of this contribution is to propose a pedagogic presentation of the theory of limit analysis (LA) and, in particular, its kinematic approach. The first domain of applications concerns the analysis of the onset of structural events, such as a thrust or a gravity collapse. The second domain is the long-term evolution of the geological structures captured with the sequential combination of LA with geometrical constructions of fault-related folds. The scheme resulting from such combination is called sequential limit analysis (SLA).

The common understanding of the shape and properties of accretionary wedges and fold-and-thrust belts relies essentially on the analytical critical Coulomb wedge (CCW) theory valid for a triangular shaped wedge of infinite extent. In a series of seminal papers (Dahlen, 1984; Davis, Suppe, & Dahlen, 1983), the topographic slope α and the decollement dip β were related to the frictional properties of the decollement and of the bulk material. A critical slope α_c, function of β and of the frictional properties, was derived. If the slope α is less than α_c, the wedge is subcritical and the compression induces internal thrusting, which increases the topographic slope to α_c. If the slope α is larger than α_c, the whole wedge glides rigidly over its decollement and is said to be supercritical. The critical slope is derived in a compressional as well as in an extensional setting as shown elegantly with a Mohr construction by Lehner (1986).

It is difficult to derive analytical solutions for applications in this geological context and most of the mechanical approaches rely on numerical methods, such as Strayer and Hudleston (1997), Stockmal, Beaumont, Nguyen, and Lee (2007), and Cruz, Malinski, Wilson, Take, and Hilley (2010). There is a multiplicity of numerical approaches and comparisons for simple wedges have been useful in understanding the differences and limitations (Buiter et al., 2006; Ellis, Schreurs, & Panien, 2004). One of the reasons for such difficulties is that the main deformation feature of accretionary wedges and fold-and-thrust belts is the faulting process; a velocity discontinuity develops in a displacement discontinuity, which are difficult to capture with any spatial interpolation, such as for the finite-element method (Ortiz, Leroy, & Needleman, 1987). The theory to detect the velocity discontinuous gradient is well established either in slip mode (Rudnicki & Rice, 1975) or in compacting band (Rudnicki, 2004). It is then possible to comprehend the competition between the onset of faulting and the onset of folding (Triantafyllidis & Leroy, 1997). The transformation of the velocity to the displacement discontinuity requires then to extend the finite-element interpolation by introducing discontinuous and singular fields (e.g., Siavelis, Guiton, Massin, & Moës, 2013).

The technical difficulty of the numerical simulations based on mechanics is at odds with the simplicity of the geometrical constructions of fault-related folds, which were developed in the structural geology literature since the seminal work of Suppe (1983). Field observations and seismic section interpretations have provided enough information to construct the geometries of most of the folds of interest without requiring a complete modeling from the mechanics point of view. The only concern is that there is no mechanics criterion to help in choosing between various geometrical constructions. This observation, shared by many (Cooke & Madden, 2014; Cooke & Murphy, 2004; Hardy, Duncan, Masek, & Brown, 1998; Masek & Duncan, 1998) prompted the two authors to introduce the concepts of minimum dissipation in a simple thrust axial surface (Maillot & Leroy, 2003). This line of work was continued by studying the development of a kink fold (Kampfer & Leroy, 2009; Maillot & Leroy, 2006) introducing the kinematic approach of LA that provides a sound mathematical framework. These studies convinced the two authors, at least, of the interest to combine LA theory and geometrical constructions and thus to develop the theory of SLA.

This contribution is an attempt to popularize the LA theory and in particular its kinematic approach. LA is a well-known theory presented in the books of Chen (1975) and Salençon (2002), and our goal here is to illustrate the potential of the method for applications to accretionary wedges and fold-and-thrust belts.

The contents of this contribution are as follows. Section 2 is devoted first to a general presentation of the LA theory and second to a detailed construction of the kinematic approach. The role of the fluid overpressures is discussed, and two simple examples are treated to illustrate the theory: the compression of a submerged block upslope and the gravitational sliding of the same block. Sections 3 and 4 are proposed to illustrate the potential of the kinematic approach of LA in structural geology and the model problem of the onset of a triangular zone at the front of an accreting wedge is proposed for sake of illustration. Such triangular zones are common, as in the Canadian Cordillera (Price, 1986; Varsek, 1996) and in the southern Pyrenees (Sans, Muñoz, & Vergés, 1996). The prototype is presented in Fig. 1 and consists of a triangular wedge accreting at its front. The triangular shape is of interest since it will help us in making the link between our predictions and the results of the CCW theory. There are four mechanical questions raised by the geometry of the triangular zone: what is the role of the delamination on the upper decollement, why is the active decollement changing its depth at the accreting layer proximity, why has the reverse fault linking the upper decollement to the free surface a listric shape, and how to explain the hinterland vergence of the upper thrusting ? A hinterland-vergent thrust is defined here by the motion of the hanging wall (HW) toward the triangular wedge. Classical thrusts are foreland-vergent, and the HW is being displaced toward the accreting layer. This question of thrust vergence is more general than for the triangular zone and is observed

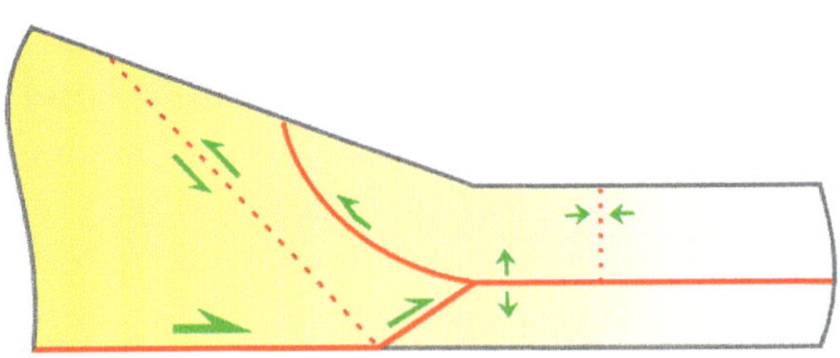

Figure 1 The idealized prototype of a triangular zone at the front of an accreting wedge with four features: the change in depth of the decollement, the delamination of the upper decollement, the curving of the ramp, and its hinterland vergence.

in Northern Sumatra and in Cascadia raising issues concerning tsunami generation after dynamic rupture (Cubas, Souloumiac, & Singh, 2016). The vergence of the thrust has certainly several possible origins and the following one is considered in this contribution; the ramp of the thrust is purely frictional whereas its conjugate axial surface is deforming in a mixed mode of compaction and shear referred to as shear-enhanced compaction (Baud, Vajdova, & Wong, 2006). The limited compaction strength (Wong, Baud, & Klein, 2001) of young sediments at the wedge front is favoring this asymmetry of the two discontinuities bounding the thrust HW, and it is proved with the LA theory that the hinterland ramp is the dominant mode of deformation. The details of the derivations for the various mechanisms discussed in the main text are postponed to the electronic supplement of this contribution, for sake of conciseness.

2. LIMIT ANALYSIS AND ITS KINEMATIC APPROACH

The objective of this section is to provide an overview of the LA theory and to concentrate on its kinematic approach, most often used in practice. For that purpose, we need some preparation and in particular the construction of the theorem of virtual power from the conditions of mechanical equilibrium. The reader unfamiliar with continuum mechanics could benefit from reading the concise presentation of Chadwick (1976) or the more developed treatment of Malvern (1969, pp. 301–320) or of Salençon (2012), the latter having exactly the same notation as in this contribution.

2.1 An Overview of Limit Analysis

LA provides bounds, typically to the tectonic force in a compressional setting. This tectonic, unknown force is represented by the vertical arrow on the left of Fig. 2. The bound is either a lower bound if the static approach of LA [blue rectangle to the right] is applied or an upper bound if the kinematic approach [red rectangle] is considered.

The static approach is certainly the easiest to explain but the most difficult to apply, at least by analytical means. The starting point is the setting up of a set of stress fields parameterized either by a couple or by a large number of scalars if a numerical approach is considered (Souloumiac, Leroy, Maillot, & Krabbenhøft, 2009). These stress fields have the fundamental property to

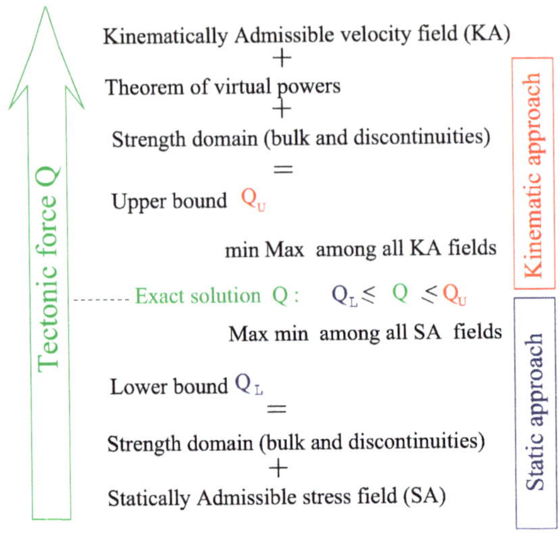

Figure 2 The theory of limit analysis is composed of two approaches, the kinematic and the static, providing upper and lower bounds to the exact, unknown tectonic forces (typical of the compressional regime).

be statically admissible (SA) meaning that they satisfy mechanical equilibrium, in the bulk material, across any discontinuity or on the boundary, three conditions which will be explained later on. A further condition is now imposed on these SA stress fields: they should satisfy pointwise the criterion that defines the maximum strength of the material composing the structure. Typically a Coulomb criterion is considered but several applications have also required to introduce a compacting threshold (Kampfer & Leroy, 2009; Maillot & Leroy, 2006). The strength domains considered in the rock mechanics literature are convex in the stress space and most often they contain the zero stress. This convexity property implies that the SA stress field that satisfies everywhere the strength criterion coincides with a force applied at the boundary is smaller than the exact, unknown tectonic force. We thus have a family of lower bounds, and now it suffices to select the free parameters as to obtain the largest lower bound Q_L. In mathematical terms, it is a *Max min* optimization problem. It is possible to construct a loading path up to that stress state without the structure collapsing. The difficulty of the method resides in selecting the parameters for the three components of the stress field in a 2D problem such that the stress field is SA and also sufficiently close to the exact stress field. Numerical solutions are, of course, more interesting for the user because of this difficulty. Please note that there is no uniqueness theorem concerning the field associated to the largest lower bound.

The static approach is not discussed any further in this contribution despite the strong interest of determining the in situ stress in active structures in the oil industry. The reader is referred to Souloumiac et al. (2009) for further discussion and applications and to Krabbenhoft and Damkilde (2003) for the algorithm currently found in the code *OptumG2* (Krabbenhøft & Lyamin, 2014) which is used most often.

The kinematic approach is the simplest of the two approaches to use, but it is also the most difficult to explain. The starting point is the construction of a collapse mechanism (CM), for example, a ramp and an axial surface bounding the HW of a fold at its onset. This mechanism corresponds to a velocity field that is virtual, first in the sense that it is a simplification of the exact field. Using the same example, the HW velocity is considered uniform with a discontinuity at the crossing of the axial surface while the footwall velocity is set to zero for sake of simplicity. This field is virtual for a second reason that its dimension (say meter per second) is irrelevant. As we will see, the velocity field is normalized by the velocity at a remote boundary where the tectonic force is applied. In other words, there is no time scale associated to the LA theory. Having constructed a set of virtual velocity fields, parameterized by a few or a large number of scalars for analytical development or numerical work, one needs to make sure that the elements of one's set are kinematically admissible (KA). A simple definition is that your velocity field is consistent with the boundary conditions of the problem. This definition will become more precise in what follows. The next step is to introduce the concept of mechanical equilibrium by application of the theorem of virtual powers, derived in the next subsection. The theorem states the equality between the external virtual power, due to the tectonic force applied on the boundary and the body forces due to gravity, and the internal virtual power for any KA velocity field. The internal virtual power is composed of the sum over the structure of the dissipation. This dissipation is defined here as the product of the stress tensor by the spatial gradient of the virtual velocity in the bulk and by the product of the stress vector acting on a discontinuity and the associated velocity jump. This equality is only of theoretical interest since in the kinematic approach, there is no information on the stress field. However, the mathematician has again recourse to convex analysis to show that the dissipation just defined has a maximum value called the support function. We will indeed see how the convexity of the strength domain leads to the selection of a virtual velocity field such that the internal virtual power has a maximum. The internal virtual power is thus bounded by the sum over the structure of the support function, a sum which is called the maximum resisting power. This new inequality is combined with the theorem of virtual powers and is then used to find an upper bound to the unknown tectonic force present in the external virtual power. This upper bound depends on the velocity field which is parameterized by a few or many scalars. Select those as to minimize the upper bound and to obtain the least upper bound. From the mathematical point of view, this is a *min Max* optimization problem. The CM associated to the least upper bound is said to be dominant for the simple reason that you cannot exceed its load without triggering it. Note again that there is no uniqueness of these dominant CMs. The best example of nonuniqueness in the context of accretionary wedges was provided by Mary, Maillot, and Leroy (2013a) who showed that under critical conditions the upper bounds for the CMs in the central region are equal up to numerical error, consistent with the CCW theory prediction.

The main difference between the LA theory summarized earlier and the solution procedure for a boundary-value problem concerns the constitutive relations. Hafner (1951) was relying on the theory of linear elasticity, which relates the stress change to the strain, to construct SA stress fields. Crans, Mandl, and Haremboure (1980) considered a pressure-sensitive plasticity theory to study growth faults, a nonlinear theory, which relates incremental stress and strain. LA accounts for the property that the stress is bounded by a strength criterion that has the essential property to be convex in stress space. However, LA does not relate the stress tensor to the virtual strain tensor or the stress vector to the virtual velocity jumps. LA does provide the means to estimate the dominant CM but does not provide the displacement field increment or displacement rates, which are typically the solution of the boundary-value problem. This limited impact is nevertheless circumvented by combining classical, geometrical, and incremental constructions of folding as proposed by Suppe (1983) and the LA theory, as it will be seen in the second part of this contribution and shown by Kampfer and Leroy (2012) for the case of a fault-propagation fold.

2.2 The Theorem of Virtual Powers

A preliminary to the presentation of the kinematic approach of LA is the recognition that the theorem of virtual powers is nothing but an integral expression of the conditions of mechanical equilibrium. This subsection is devoted to the derivation of this theorem and is illustrated with the simple example of a composite block standing on an inclined ramp, Fig. 3A.

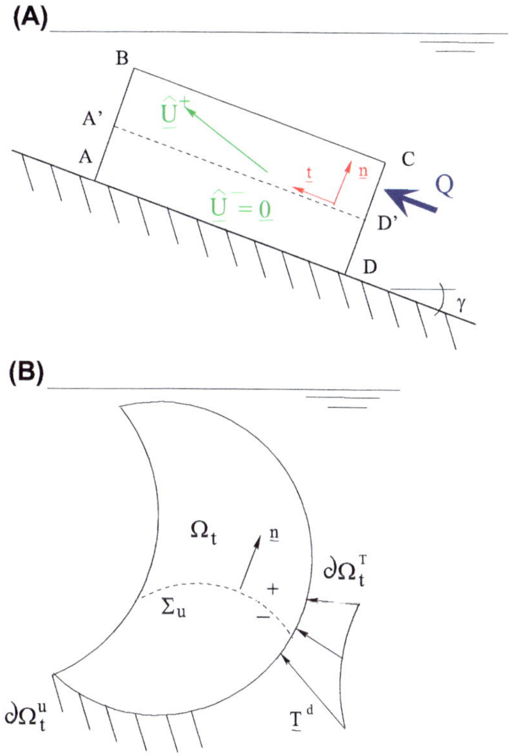

Figure 3 The simple example of a submerged composite block hanging on an inclined ramp is proposed to illustrate the kinematic approach of limit analysis (A) and (B) the general notation for any structure.

The block is composed of two layers with contact along A′D′, composed of different materials. The segment AD is glued on the inclined ramp, and if collapse occurs, it corresponds to a slip along the interface A′D′. The unknown tectonic force applied on the surface D′C has the norm Q and is parallel to the sliding surface. More generally, this composite block corresponds to the domain Ω_t of boundary $\partial\Omega_t$, the time t identifying the current position of the domain. This boundary is decomposed into two parts, $\partial\Omega_t^u$ and $\partial\Omega_t^T$ corresponding to AD and DC, respectively, in our example. On the former surface, the displacement is known to be $\underline{u}^d$ (zero in our particular example) and on the latter surface, the force density $\underline{T}^d$ (force per unit area) is prescribed. Note the discontinuity surface Σ_u in Fig. 3B, which corresponds to A′D′ in our example.

The local conditions for mechanical equilibrium are obtained from the principle of statics, which states that the sum of all the external forces and the sum of the moments of the external forces with respect to any point are zero. One derives from this principle (Chadwick, 1976; Salençon, 2012), the three conditions

$$\underline{\mathrm{div}}(\underline{\underline{\sigma}}) + \rho \underline{g} = \underline{0} \quad \forall \underline{x} \in \Omega_t, \tag{1}$$

$$[\![\underline{\underline{\sigma}}]\!] \cdot \underline{n} = \underline{0} \quad \forall \underline{x} \in \Sigma_U,$$

$$\underline{T} = \underline{T}^d \quad \forall \underline{x} \in \partial \Omega_t^T.$$

Quantities underlined once and twice are vectors and second-order tensors, respectively. The first equation in (1) means that the divergence of the stress tensor plus the gravity effects (ρ and $\underline{g}$ are the volumetric mass and the gravity acceleration) is a zero vector at any point within the domain Ω_t. The second equation expresses the same concept across the discontinuity Σ_u oriented by its normal $\underline{n}$. The double brackets mean the difference between the stress tensor on the positive and the negative sides of the discontinuity, the former being pointed by the normal $\underline{n}$. The third equation states that the stress vector $\underline{T}$ on the boundary $\partial\Omega_t^T$ must be equal to the applied force density $\underline{T}^d$ (force per unit area).

Consider now a virtual velocity field $\underline{\widehat{U}}$, piecewise continuous on the two subdomains separated by Σ_u. The superposed hat on the velocity is to stress the virtual property of the field. Multiply the first equation in (1) by this velocity and sum the resulting scalar over one of the subdomains denoted by $\Omega_t^{'}$

$$\int_{\Omega_t^{'}} \underline{\widehat{U}} \cdot \left(\underline{\mathrm{div}}(\underline{\underline{\sigma}}) + \rho \underline{g} \right) dV = 0, \tag{2}$$

$$\int_{\Omega_t^{'}} \widehat{U}_i \left(\sigma_{ij,j} + \rho g_i \right) dV = 0.$$

The second equation in (2) is equivalent to the first mentioned previously and is written with an indicial notation to facilitate the introduction of the tensorial notation. A comma stands for partial differentiation with respect to the coordinate (x_j here), and repeated indices signal a summation from 1 to 3. Similar repetitions of the equations are proposed in what follows for the same purpose. Eq. (2) is now integrated by parts

$$\int_{\Omega_t^{'}} \mathrm{div}\left(\underline{\widehat{U}} \cdot \underline{\underline{\sigma}} \right) - \underline{\underline{\nabla \widehat{U}}} : \underline{\underline{\sigma}} + \underline{\widehat{U}} \cdot \rho \underline{g} \; dV = 0, \tag{3}$$

$$\int_{\Omega_t'} \left(\widehat{U}_i\sigma_{ij}\right)_{,j} - \widehat{U}_{i,j}\sigma_{ij} + \widehat{U}_i\rho g_i \, dV = 0.$$

The divergence theorem is applied on the first term of the integrand, and Eq. (3) becomes

$$\int_{\partial\Omega_t'} \underline{\widehat{U}} \cdot \underline{\underline{\sigma}} \cdot \underline{n} \, dS + \int_{\Omega_t'} \underline{\widehat{U}} \cdot \rho \underline{g} \, dV = \int_{\Omega_t'} \underline{\underline{\nabla \widehat{U}}} : \underline{\underline{\sigma}} \, dV, \quad (4)$$

in which $\underline{n}$ is the unit normal on the boundary pointing to the exterior. Recall that the principle of statics leads to the proof that the stress tensor is symmetric, and note that the double product $\underline{\underline{\nabla \widehat{U}}} : \underline{\underline{\sigma}}$ is equal to $\underline{\underline{d}}(\underline{\widehat{U}}) : \underline{\underline{\sigma}}$ in terms of the rate of deformation tensor $\underline{\underline{\widehat{d}}} = 1/2({}^t\underline{\underline{\nabla \widehat{U}}} + \underline{\underline{\nabla \widehat{U}}})$, which is the symmetric part of the virtual velocity gradient.

The virtual velocity field is now constrained such that it is zero on the boundary $\partial\Omega_t^{u'}$ where the displacement is prescribed (the segment AD in the example). Furthermore, on the remaining part of the boundary, the Cauchy relation ($\underline{T} = \underline{\underline{\sigma}} \cdot \underline{n}$) and the third equilibrium condition in Eq. (1) are applied. Let us assume for the time being and for sake of simplicity that the contact region Σ_u does not exist. These statements lead to the new expression of Eq. (4):

$$\int_{\partial\Omega_t^{T'}} \underline{\widehat{U}} \cdot \underline{T}^d \, dS + \int_{\Omega_t'} \underline{\widehat{U}} \cdot \rho \underline{g} \, dV = \int_{\Omega_t'} \underline{\underline{d}}(\underline{\widehat{U}}) : \underline{\underline{\sigma}} \, dV. \quad (5)$$

The left part of this equality corresponds to the virtual power of the external forces, either body force or applied force density on the boundary. It is denoted by $\mathscr{P}_{\text{ext}}$. The right part represents the internal power, $\mathscr{P}_{\text{int}}$, done by the stress field on the virtual rate of deformation tensor. Eq. (5) states the equality between the external and the internal virtual powers for any KA virtual velocity:

$$\mathscr{P}_{\text{ext}} = \mathscr{P}_{\text{int}} \quad \forall \underline{\widehat{U}} \quad KA. \quad (6)$$

We now propose to extend this theorem of virtual power to domains containing a velocity discontinuity Σ_U. We have disregarded so far the contribution of such discontinuities in the external power in Eq. (5). The new additional term corresponds to the integral over Σ_u' of the product $\underline{T}' \,.\, \underline{U}'$ (the prime corresponds either to the positive side or the negative side of the discontinuity). Note that $\underline{T}^+ = \underline{\underline{\sigma}}^+ \cdot \underline{n}^+ = -\underline{\underline{\sigma}}^+ \cdot \underline{n} = -\underline{\underline{\sigma}}^- \cdot \underline{n} = -\underline{\underline{\sigma}}^- \cdot \underline{n}^- = -\underline{T}^-$ by application of the Cauchy relation, of the third equilibrium condition in Eq. (1) and defining the normals $\underline{n}^\pm$

to be the normal pointing to the exterior of the plus and minus subdomain, respectively. These remarks are now put into use as it is proposed to apply the theorem of virtual power (5) on the two subdomains of common boundary Σ_u, and to add the two equations resulting in

$$\int_{\partial\Omega_t^T} \underline{\widehat{U}} \cdot \underline{T}^d dS + \int_{\Omega_t} \underline{\widehat{U}} \cdot \rho \underline{g} \, dV + \int_{\Sigma_u} \left(\underline{\widehat{U}}^+ \cdot \underline{T}^+ + \underline{\widehat{U}}^- \cdot \underline{T}^- \right) dS$$
$$= \int_{\Omega_t} \underline{\underline{d}}\left(\underline{\widehat{U}}\right) : \underline{\underline{\sigma}} \, dV. \tag{7}$$

Replace now the stress vector $\underline{T}^+$ on Σ_u by $-\underline{T}^-$ and the integral over this common interface becomes minus the stress vector times the velocity jump ($[\![\underline{\widehat{U}}]\!] = \underline{\widehat{U}}^+ - \underline{\widehat{U}}^-$), a contribution which is moved to the right-end side of Eq. (7) so that the virtual internal power now reads

$$\mathscr{P}_{\text{int}} = \int_{\Omega_t} \underline{\underline{d}}\left(\underline{\widehat{U}}\right) : \underline{\underline{\sigma}} \, dV + \int_{\Sigma_u} [\![\underline{\widehat{U}}]\!] \cdot \underline{T} \, dS. \tag{8}$$

This extension is essential for what follows since most of the examples treated the virtual velocity field in piecewise uniform. Consequently, the gradient of the velocity is zero and so is also its symmetric part, the rate of deformation tensor. The only contribution to the internal virtual power is then due to the velocity jump now present in the definition of the internal power in Eq. (8).

Two additional developments are necessary in this preliminary subsection, and it concerns the fluid-saturated aspect of the continuum at hand. The first development concerns the choice of stress. In what follows the strength criterion is based on Terzaghi effective stress $\underline{\underline{\sigma}}' = \underline{\underline{\sigma}} + p_f \underline{\underline{\delta}}$ in which p_f is the fluid pressure and $\underline{\underline{\delta}}$ is the second-order identity tensor. The stress vector follows a similar decomposition with the introduction of the effective stress vector $\underline{T}' = \underline{T} + p_f \underline{n}$. Note that the effective stress is denoted with a prime which has nothing to do with the prime used previously to differentiate between the two subdomains. Inserting these two decompositions in Eq. (8), we define the internal, effective virtual power by

$$\mathscr{P}'_{\text{int}} = \int_{\Omega_t} \underline{\underline{d}}\left(\underline{\widehat{U}}\right) : \underline{\underline{\sigma}}' \, dV + \int_{\Sigma_u} [\![\underline{\widehat{U}}]\!] \cdot \underline{T}' \, dS. \tag{9}$$

The contribution of the fluid pressure is shifted to the external power which becomes the external, effective virtual power

$$
\begin{aligned}
\mathscr{P}'_{\text{ext}} = & \int_{\partial\Omega_t^T} \underline{\widehat{U}} \cdot \underline{T}^d dS + \int_{\Omega_t} \underline{\widehat{U}} \cdot \rho \underline{g} \, dV + \int_{\Omega_t} p_f \operatorname{tr}\left(\underline{\underline{\widehat{U}}}\right) dV \\
& + \int_{\Sigma_u} p_f [\![\underline{\widehat{U}}]\!] \cdot \underline{n} \, dS,
\end{aligned}
\tag{10}
$$

in which $\operatorname{tr}(\underline{\underline{\widehat{U}}})$ is the trace of the rate of deformation tensor or the virtual rate of volume change, and $[\![\underline{\widehat{U}}]\!] \cdot \underline{n}$ represents the virtual opening rate of the discontinuity. Mechanical equilibrium is then presented through the following theorem of effective virtual powers:

$$
\mathscr{P}'_{\text{ext}} = \mathscr{P}'_{\text{int}} \quad \forall \underline{\widehat{U}} \quad KA. \tag{11}
$$

The second development concerns the fluid pressure on the boundary $\partial\Omega_t^{T'}$ of a region Ω_t'. The fluid exerts a pressure p_f that is oriented along the normal to the boundary. If that boundary is only subjected to the hydrostatic fluid pressure p_h, Archimedes' theorem states

$$
\int_{\partial\Omega_t^{T'}} - p_h \underline{n} \, dS = - \int_{\Omega_t'} \rho_f \underline{g} \, dV, \tag{12}
$$

in which ρ_f is the fluid volumetric mass. Consider now a uniform velocity field over this subdomain and multiply the two sides of Eq. (12) by this velocity to obtain

$$
\int_{\partial\Omega_t^{T'}} - p_h \underline{n} \cdot \underline{\widehat{U}} \, dS + \int_{\Omega_t'} \rho_f \underline{g} \cdot \underline{\widehat{U}} \, dV = 0, \tag{13}
$$

a useful result to simplify many problems with a complex topography. It is the fluid pressure in excess of the hydrostatic pressure, Δp_f, called the overpressure, which is then introduced to benefit from this simplification. This theorem is often referred to as the weak form (integral form) of Archimedes' theorem.

2.3 The Kinematic Approach

2.3.1 The Collapse Mechanism and the Virtual Powers

The starting point of the kinematic approach is the construction of a CM and the associated KA displacement field. The CM for the simple block hanging on the inclined ramp in Fig. 3 is the sliding of the upper part of the block over the interface A′D′. The associated velocity field is such that the virtual velocity is zero in the lower layer and equal to the vector $\underline{\widehat{U}}^+$ in the upper layer. This virtual velocity field is KA since it is zero on

AD where the displacements are prescribed. Note that the norm and the direction of $\underline{\widehat{U}}^+$ are yet undefined, introducing two degrees of freedom in the problem.

The second step consists in applying the theorem of effective virtual powers (Eq. 11) together with the weak form of Archimedes' theorem (Eq. 13) to the problem at hand. The effective external power reads

$$\mathscr{P}'_{\text{ext}} = \underline{\widehat{U}}^+ \cdot \underline{t} Q' + \underline{\widehat{U}}^+ \cdot \left(\rho - \rho_f\right) \underline{g} S_{A'BCD'} + \underline{\widehat{U}}^+ \cdot \underline{n} \int_{\Sigma_u} \Delta p_f dS. \quad (14)$$

The first of the three contributions is the power of the velocity on the effective tectonic force (there is a contribution of the hydrostatic pressure), which is acting parallel to the interface A′D′ (unit tangent vector $\underline{t}$). The second contribution concerns the power of the velocity on the gravity forces, the scalar $S_{A'BCD'}$ denoting the volume of the upper layer. The third contribution is the work of the opening component of the velocity field ($\underline{n}$ is the unit normal to A′D′) on the fluid overpressure. This difference in pressure is assumed to be known. The effective, virtual internal power is

$$\mathscr{P}'_{\text{int}} = \underline{\widehat{U}}^+ \cdot \int_{\Sigma_u} \underline{T}' dS. \quad (15)$$

There is a single dissipation source, which is the power of the velocity jump on the effective traction vectors acting on the interface Σ_u. There is no information on the stress state within the structure, so it is impossible to estimate the effective, virtual internal power and thus to deduce the tectonic force Q in the effective, external virtual power. Nevertheless, the LA theory provides a bound to this force by recognizing that the stress vector acting on the interface is constrained by its limited strength.

2.3.2 The Support Function

The effective stress vector ($\underline{T}' = \underline{\underline{\sigma}}' \cdot \underline{n}$) is decomposed in effective normal and tangential components

$$\underline{T}' = \sigma'_n \underline{n} + \tau \underline{t}, \quad (16)$$

in the right-handed basis $(\underline{n}, \underline{t})$ attached to the discontinuity [in red in Fig. 3]. In the two-dimensional stress space (σ'_n, τ), if the general criterion $f(\sigma'_n, \tau)$ is strictly negative then the discontinuity remains inactive. Slip requires the criterion to be attained: $f(\sigma'_n, \tau) = 0$. Typically, this criterion is the Coulomb criterion but to remain general for the time being, consider the elliptical shaped criterion presented in Fig. 4A. This general criterion contains the origin of the stress space and also is convex. It means that if the two sets (σ'_{n1}, τ_1) and (σ'_{n2}, τ_2) are within or on the boundary of the domain G bounded by the condition $f(\sigma'_n, \tau) = 0$, then any linear combination $(\sigma'_{n1}, \tau_1)(1 - \lambda) + (\sigma'_{n2}, \tau_2)\lambda$, for a scalar $\lambda \in [0;1]$, is also within or on the boundary of the domain G. The elliptical criterion of Fig. 4A has this property. The reader will check that the Coulomb criterion, despite the fact that the cone is open in compression also has this convexity property.

This convexity property is essential to derive the bound to the dissipation, and a graphical method is proposed to compute it. Fig. 4B presents a velocity discontinuity, the dashed segment, and its normal and tangent vectors are oriented as in Fig. 4A; the effective normal stress is along the normal direction and the tangential stress is parallel to the tangent vector. The jump in velocity $[\![\underline{\widehat{U}}]\!]$ results from the difference between the velocities on the plus and minus sides of the discontinuity. This velocity jump is normal to a plane (line), which is presented as a green dashed segment. Introduce this plane, conserving its orientation with the stress space and displace it toward the elliptical strength domain until contact occurs. The stress vector at this contact point is denoted by $\underline{T}^*$.

We are going to establish that the maximum dissipation $\underline{T}' \cdot [\![\underline{\widehat{U}}]\!]$ for all possible effective stress vectors $\underline{T}'$ touching the boundary of the domain G is $\underline{T}^* \cdot [\![\underline{\widehat{U}}]\!]$, also denoted $\pi([\![\underline{\widehat{U}}]\!])$ and referred to as the support function:

$$\mathrm{Sup}_{G(\underline{T}')} \underline{T}' \cdot [\![\underline{\widehat{U}}]\!] = \underline{T}^* \cdot [\![\underline{\widehat{U}}]\!] = \pi([\![\underline{\widehat{U}}]\!]), \tag{17}$$

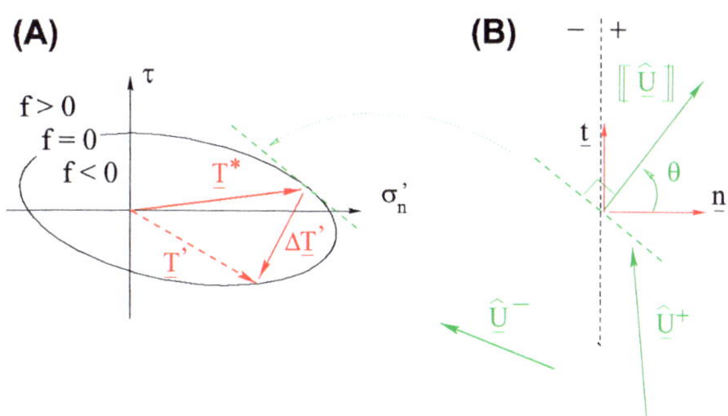

Figure 4 Construction of the support function for an elliptically shaped strength domain in the tangential and normal stress space (A). In (B) the velocity jump across the discontinuity (*vertical dashed segment*) is defined as the difference between the velocities on the plus and the minus sides.

For this, consider any $\underline{T}'$, which is equal to $\underline{T}^* + \Delta\underline{T}'$, as illustrated in Fig. 4A. The product $\underline{T}' \cdot [\![\underline{\widehat{U}}]\!]$ is equal to $\pi([\![\underline{\widehat{U}}]\!])$ plus $\Delta\underline{T}' \cdot [\![\underline{\widehat{U}}]\!]$. The second product is null or negative since $\Delta\underline{T}'$ is pointing in the direction opposite to the normal to the dashed line tangent to G, which is oriented by the velocity jump. We can thus conclude that the support function is a bound to the product $\underline{T}' \cdot [\![\underline{\widehat{U}}^+]\!]$ as stated in Eq. (17). This reasoning is facilitated by the simple two-dimensional character of the stress space, but it can be generalized to the full six-dimensional stress space of continuum mechanics for diffuse deformation (Salençon, 2002) using instead of the velocity jump the rate of deformation tensor ($\underline{\underline{\sigma}}'$ and $\underline{\underline{\widehat{d}}}$).

It is worth to apply this geometrical construction of the support function to the Coulomb criterion, presented in Fig. 5. It is presented as an open cone in the compressive direction of internal angle twice the friction angle ϕ. The tip of the cone is in the positive (extension) direction because of the introduction of a cohesion C. Let us place in the same stress space the velocity jump oriented by the angle θ from the normal (see Fig. 4) and move the dashed plane toward the cone. Three cases are possible, labeled cases 1–3, based on the orientation of θ, considered positive counterclockwise. If θ is less than $\pi/2 - \phi$, this is case 1 and the stress vector $\underline{T}^*$ is presented in red since the contact point is the tip of the cone. If the angle θ is exactly $\pi/2 - \phi$, case 2, the stress vector $\underline{T}^*$is any vector touching the cone, as shown with the blue vector. In case 3, the angle θ is exceeding $\pi/2 - \phi$ and the stress vector $\underline{T}^*$ is infinite. Having estimated the vector $\underline{T}^*$ for all positive θ, the support function is computed and summarized by

$$\begin{aligned} &\text{case}(1,1') : 0 \leq |\theta| < \pi/2 - \phi, \quad \pi\left([\![\underline{\widehat{U}}]\!]\right) = \widehat{J} C \,\text{cotan}(\phi)\cos\theta, \\ &\text{case}(2,2') : |\theta| = \pi/2 - \phi, \quad \pi\left([\![\underline{\widehat{U}}]\!]\right) = \widehat{J} C \cos\phi, \\ &\text{case}(3,3') : \pi/2 - \phi < |\theta| \leq \pi, \quad \pi\left([\![\underline{\widehat{U}}]\!]\right) = +\infty, \end{aligned} \tag{18}$$

where $\widehat{J}$ is the norm of the $[\![\underline{\widehat{U}}]\!]$. Note that a similar construction is done in Fig. 5 for negative θ and corresponding to cases 1′–3′. The support function is the same regardless of the sign of this angle.

A word of caution is required for case 3 which includes the expected solution by a field geologist, namely $\theta = \pi/2$ for which the velocity jump is tangent to the fault. This special choice and any selection according to case 3 lead to an infinite bound to the dissipation. Such infinite bound

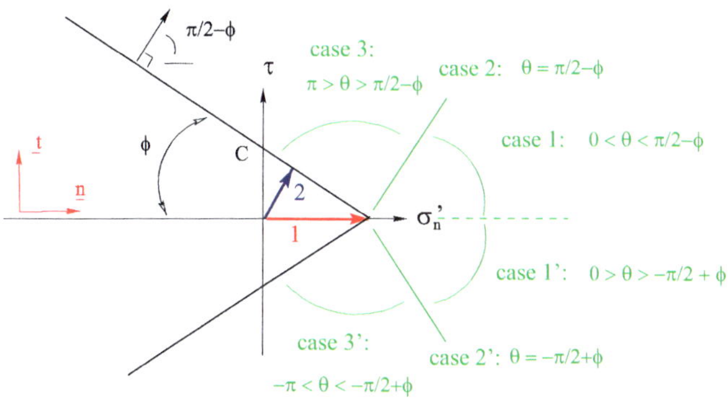

Figure 5 Construction of the support function for the Coulomb criterion. There are three cases in the orientation of the velocity jump with respect to the normal to the discontinuity, marked by the angle θ.

has no merits for the proposed methodology and should be avoided. KA virtual fields that avoid this pitfall are said to be pertinent.

This geometrical construction of the support function is conducted for a more general strength criterion, Fig. 6. It complements the Coulomb criterion by the addition of the collapse pressure in a hydrostatic test P^* and of a linear cap for shear-enhanced compaction characterized by the angle ϕ^*. In addition, a tensile cut-off is introduced with the tensile strength T. The geometrical construction of the support function provides six cases noted from 0 to 5, so that cases 1–3 are common with the classical Coulomb criterion. Case 0 corresponds to a pure opening of the discontinuity. Case 3 is now of interest and corresponds to the cone made by the two linear segments for the frictional and the compacting mechanisms. Case 4 is the shear-enhanced mode of deformation. Case 5 is the cone between a pure compaction and the shear-enhanced compaction, typical of a compaction band type of instability. In summary, the generalized support function reads

$$\text{case}(0): \theta = 0, \quad \pi(\underline{\widehat{J}}) = \widehat{J}T, \tag{19}$$

$$\text{case}(1): 0 < \theta < \pi/2 - \phi, \quad \pi(\underline{\widehat{J}}) = \widehat{J}\left[\frac{\cos(\theta+\phi)}{\cos(\phi)}T + C\sin(\phi)\right],$$

$$\text{case}(2): \theta = \pi/2 - \phi, \quad \pi(\underline{\widehat{J}}) = \widehat{J}C\cos(\phi),$$

$$\text{case}(3): \pi/2 - \phi < \theta < \pi/2 + \phi^*, \quad \pi(\underline{\widehat{J}}) = \widehat{J}[\cos(\theta)\overline{\sigma} + \sin(\theta)\overline{\tau}],$$

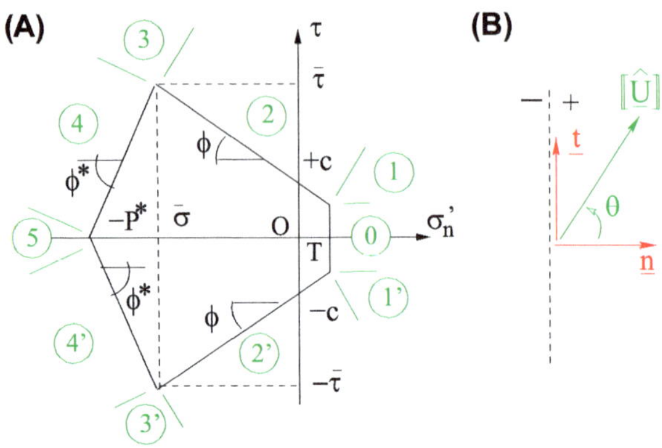

Figure 6 A more general strength criterion than Coulomb criterion by introduction of a cap in the compressive range as well as a cut-off in tension (A). The orientation θ of the velocity jump with respect to the normal of the discontinuity leads to six cases for the construction of the support function numbered from 0 to 5 (B).

$$\text{case}(4) : \theta = \pi/2 + \phi^*, \quad \pi(\underline{\hat{J}}) = \hat{J}P^* \sin(\phi^*),$$

$$\text{case}(5) : \pi/2 + \phi^* < \theta < \pi, \quad \pi(\underline{\hat{J}}) = \hat{J}P^* \cos(\theta),$$

in which the two scalars introduced for case 3 are

$$\bar{\sigma} = \frac{C - P^* \tan\phi^*}{\tan\phi + \tan\phi^*}, \quad \bar{\tau} = \frac{(C + P^* \tan\phi)\tan\phi^*}{\tan\phi + \tan\phi^*}. \tag{20}$$

They correspond to the intersection of the Coulomb line and the compaction truncation line presented in Fig. 6A. Note that Eq. (19) was written for θ in the range $[0;\pi]$ but the symmetry in Fig. 6 with respect to the effective normal stress axis points to the symmetric cases $1'-5'$ for negative angle θ in the range $[0;-\pi]$.

2.3.3 The Least Upper Bound

We have just shown that the maximum dissipation is given by the support function. Let us sum this function over the discontinuities of type Σ_u and obtain the maximum resisting power which is larger than the effective virtual internal power

$$\mathscr{P}'_{\text{int}}(\underline{\hat{U}}) \leq \mathscr{P}'_{\text{mr}}(\underline{\hat{U}}) \quad \text{with} \quad \mathscr{P}'_{\text{mr}}(\underline{\hat{U}}) = \int_{\Sigma_u} \pi([\![\underline{\hat{U}}]\!])\, dS. \tag{21}$$

It is worth applying this result to our simple example of Fig. 3 with a ramp dipping at γ. The interface A′D′ is assumed to be cohesive and frictional and to obey the Coulomb criterion such that the support function (Eq. 18) does apply. The effective, external virtual power in Eq. (14) is equal to the unknown internal power, which is less than the maximum resisting power according to Eq. (21) leading to

$$\begin{aligned}\widehat{J}\sin(\theta)Q' - \widehat{J}\cos(\gamma-\theta)\left(\rho-\rho_f\right)gS_{A'BCD'} + \widehat{J}\cos(\theta)\Delta\overline{p}_f L_{A'D'}\\ \leq \widehat{J}CL_{A'D'}\cos(\theta)\operatorname{cotan}(\phi),\end{aligned} \tag{22}$$

having replaced the velocity $\underline{\widehat{U}}^+$ by the norm of the velocity jump and its orientation θ from the normal to the discontinuity. In Eq. (22), we have also introduced the average fluid overpressure $\Delta\overline{p}_f$ along A′D′ and accounted for the support function to be according to case 2. Note that in this equation $L_{A'D'}$ and $S_{A'BCD'}$ denote the length of the contact $A'D'$ and the surface of the top layer, respectively.

The first of two remarks is that the two sides of Eq. (22) can be normalized by the velocity jump norm, which has thus no bearing on the bound calculation. The second remark is that we have then obtained an upper bound for the effective tectonic force Q'

$$\begin{aligned}Q' \leq Q'_u(\theta) = \frac{\cos(\gamma-\theta)}{\sin(\theta)}\left(\rho-\rho_f\right)gS_{A'BCD'}\\ + \operatorname{cotan}(\theta)\Big(-\Delta\overline{p}_f + C\operatorname{cotan}(\phi)\Big)L_{A'D'}.\end{aligned} \tag{23}$$

The upper bound is a function of the angle θ defining the orientation of the virtual velocity. It has three contributions, the first due to the effective weight of the submerged structure, the second due to the fluid overpressure within the interface, and the third due to the cohesive strength of the interface.

The last step concerns the selection of the velocity orientation θ to minimize the upper bound in Eq. (23). In our example, the intuitive feeling that the velocity should be as parallel as possible from the ramp will be checked by a simple trigonometric calculation leading indeed to the conclusion that θ should be chosen as $\pi/2 - \phi$ and corresponding to case 2 for the support function in Eq. (18). The least upper bound to the effective tectonic force is thus

$$Q'_{lu} = \frac{\sin(\gamma+\phi)}{\cos(\phi)}\left(\rho-\rho_f\right)gS_{A'BCD'} + \Big(-\Delta\overline{p}_f\tan\phi + C\Big)L_{A'D'}. \tag{24}$$

Note that case 2 is often the optimum orientation and the velocity field in many applications is constructed directly with this assumption. An example with a proof is found in the electronic supplement of Cubas, Leroy, and Maillot (2008).

2.3.4 Application to Gravity Collapse

The example used throughout this section is concerning a compressive loading and the whole approach is geared to the derivation of an upper bound to the tectonic force. The approach is theoretically the same for gravitational instability although the bound concerns a geometrical parameter (e.g., the height of a cliff) or a fluid pressure. To illustrate the idea, let us keep with the example of Fig. 3 with the following two differences: there is no tectonic forces Q and the velocity vector $\underline{U}^+$ is now directed downslope. Compute the new external power according to Eq. (10) and choosing the angle θ to be in the range $[-\pi/2+\phi;0]$ select the support function in case 1′ or 2′ so that the application of the kinematic approach of LA leads to

$$-\widehat{J}\cos(\gamma-\theta)\left(\rho-\rho_f\right)gS_{A'BCD'}+\widehat{J}\cos(\theta)\Delta\overline{p}_f\,L_{A'D'} \leq \widehat{J}CL_{A'D'}\cos(\theta)\mathrm{cotan}(\phi). \tag{25}$$

Normalize Eq. (25) by the velocity jump norm and obtain the upper bound to the fluid overpressure

$$\Delta\overline{p}_f\leq\Delta\overline{p}_{fu}=\frac{\cos(\gamma-\theta)}{\cos(\theta)}\left(\rho-\rho_f\right)g\frac{S_{A'BCD'}}{L_{A'D'}}+C\,\mathrm{cotan}(\phi). \tag{26}$$

The last task is to minimize the upper bound which is done by selecting the angle $\theta=-\pi/2+\phi$ corresponding to case 2′ of the support function. The least upper bound is thus

$$\Delta\overline{p}_{flu}=\frac{\sin(\phi-\gamma)}{\sin(\phi)}\left(\rho-\rho_f\right)g\frac{S_{A'BCD'}}{L_{A'D'}}+C\,\mathrm{cotan}(\phi). \tag{27}$$

This is the maximum overpressure that the interface A′D′ can sustain without the upper block starting to slide downslope.

The influence of the fluid overpressure is not discussed in the rest of this contribution. The interested reader is referred to Pons and Leroy (2012) and to Yuan, Leroy, and Maillot (2015) for further discussion of the theory and for applications in the compressive and the extensional context, respectively. An application that accounts for the fluid overpressure to understand the activation of splay faults in Nankai wedge, Japan, was considered by Pons, Leroy, and Lallemant (2013).

3. TRIANGULAR WEDGE: SIX COLLAPSE MECHANISMS IN COMPETITION

One of the advantages of the kinematic approach of LA is that several CMs can be compared to determine the dominant mode of deformation. It is the objective of this section to compare six CMs for a triangular wedge of finite extent and of taper angle $\alpha + \beta$, the sum of the dips of the decollement and the topography. Apart from the finite extent, this wedge geometry is the one considered for the CCW theory. It will thus be possible to show how the kinematic approach reproduces the results of the CCW and opens up new horizons.

The bulk material is assumed cohesive, frictional, and compacting, as defined by the general strength domain depicted in Fig. 6. The bulk tensile strength is disregarded. The decollement is assumed to be cohesive and frictional with a limited tensile strength T. Its compaction strength is assumed infinite ($P^* \to +\infty$), so the strength domain is open in compression as for the classical Coulomb criterion. The motivation for proposing each mechanism is reviewed and the associated upper bound Q_u^i is provided. The details of the calculations are postponed to the electronic supplement.

3.1 Collapse Mechanisms 1 and 2: For Comparison With the Critical Coulomb Wedge Theory

The CM 1 is the simplest and should be dominant for supercritical wedges if only friction prevails. It consists of the gliding of the whole wedge on the decollement, Fig. 7A, where the tectonic force applied on AC has not been represented, as in the other five sketches. Note also that the coordinate system is attached to the decollement (first direction) so gravity acts at the angle β from the x_2 coordinate direction. The virtual velocity field is uniform and is oriented according to case 2 in Eq. (19), to be optimum. This choice was guided by the result of the example treated in the previous section. The kinematic approach of LA provides the upper bound to the applied force

$$Q_u^{(1)} \cos(\phi_D) = C_D \cos\phi_D L_{AB} + \rho g \sin(\phi_D + \beta) S_{ABC}, \tag{28}$$

where L_{AB} and S_{ABC} stand for the total length of the decollement and the surface of the wedge, respectively. In what follows, subscripts are sets of two

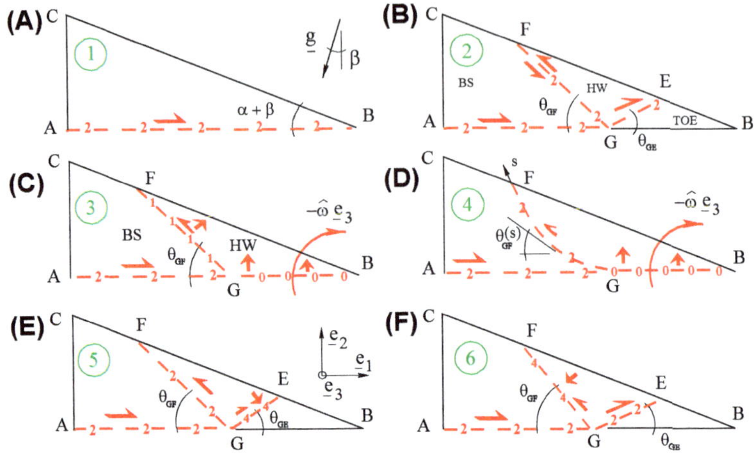

Figure 7 The six collapse mechanisms in competition in a triangular wedge: the complete gliding on the decollement (A), the classical frictional thrust fold for comparison with the CCW theory (B), the delamination mechanism with a straight ramp (C), and a curvy ramp (D), and finally the compacting thrust fold to promote hinterland (E), and foreland vergence (F). The red numbers 0, 2, and 4 corresponds to the cases of the general strength domain defined in Fig. 6 and thus identify the orientation of the velocity jump.

or three points or one or two letters that identify the lengths, the surfaces, or the material properties (D for the decollement in Eq. (28)). Note that the upper bound is not a function of any free variable since the velocity vector orientation has been already selected. Consequently, Eq. (28) is also the least upper bound $Q_{lu}^{(1)}$.

CM 2 corresponds to the onset of a thrust fold that partitions the wedge into three regions (Fig. 7B). The region GEF is the HW, which is moving forward over the fault GE. Material in the region $AGFC$, called the backstop (BS), is gliding over the decollement activated between points A and G. The toe region GBE is at rest with respect to the observer. The discontinuities GE and GF are called tentatively the ramp and the axial surface, respectively. The dip of the discontinuity designated by the two letters XY is denoted θ_{XY}, a notation which applies throughout this contribution.

Note that the number on the discontinuities of each CM presented in Fig. 7 corresponds the orientation of the virtual velocity jump defined by six cases for the general support function in Eq. (19). The velocity field is piecewise uniform and is presented in detail in Cubas et al. (2008) and in the electronic supplement. The BS is sliding on the decollement with the same velocity as for CM 1: its orientation corresponds to case 2 of the support function but its magnitude is not defined. The velocity of the HW is oriented in a similar way with respect to the ramp *GE*. There are thus two unknowns, corresponding to the magnitude of the two velocities. One is determined by imposing that the velocity jump on the axial surface *GF* is also oriented according to case 2: application of the law of sines on the hodogram construction, found in the ES, defining this velocity jump provides

$$\frac{\widehat{U}_{HW}}{\sin(\phi_D + \phi_{GF} + \theta_{GF})} = \frac{\widehat{J}_{GF}}{\sin(\phi_{GE} + \theta_{GE} - \phi_D)} = \frac{1}{\sin(\theta_{GF} + \theta_{GE} + \phi_{GF} + \phi_{GE})}. \tag{29}$$

The remaining unknown is also a velocity norm, the velocity norm of the sliding block. It is simply immaterial and used for normalization, as it is done in Eq. (29). The final result for the upper bound for this CM 2 is

$$\begin{aligned} Q_u^{(2)} \cos(\phi_D) = \rho g\big(S_{BS} \sin(\phi_D + \beta) + \widehat{U}_{HW} S_{HW} \sin(\phi_{GE} + \theta_{GE} + \beta)\big) \\ + L_{AG} C_D \cos \phi_D + C_{GF} L_{GF} \widehat{J}_{GF} \cos \phi_{GF} \\ + C_{GE} L_{GE} \widehat{U}_{HW} \cos \phi_{GE}. \end{aligned} \tag{30}$$

The upper bound in Eq. (30) is a function of three parameters, the two dips θ_{GE} and θ_{GF} as well as the distance L_{AG}. It is the minimum value of Eq. (30) that is of interest and the associated optimum values of these three parameters. The minimum load is then the least upper bound. For that minimization, we follow Cubas et al. (2008) and discretize the topography and the decollement and select the optimum position of the three points *E*, *F*, and *G* (see Fig. 7B for definition).

Note that the vergence of the thrust following a shortening of a finite increment has not been identified here. The two faults *GE* and *GF*, despite the proposed interpretation as ramp and axial surface, respectively, play identical role in this theory. They can act indeed either as a ramp and shear plane (foreland vergence) or as a shear plane and a ramp (hinterland vergence) or finally as two ramps (pop-up structure).

This second mechanism is essential to understand the critical taper theory for a purely frictional material. It is seen in what follows that the distance L_{AG} is small and thus the thrust fold at the back for subcritical conditions

($\alpha < \alpha_c$) and equal to the distance L_{AB}, for supercritical conditions ($\alpha > \alpha_c$) since the deformation is then to the front. In the latter case, the first and second CMs are identical.

3.2 Collapse Mechanism 3: Delamination on the Frontal Part of the Decollement

This CM 3 is the first to be proposed for exploring the possibility for the frontal part of the decollement to delaminate, a basic mechanism observed in most triangular zones, Fig. 7C. The wedge is partitioned into two regions, the first region being the BS $AGFC$ gliding on the section AG of the decollement. The second region is the hanging wall GBF which sustains a rigid rotation $\widehat{\omega}$ around the point B, as illustrated in Fig. 7C.

The decollement between points A and G is sheared whereas points between G and B have a velocity normal to the decollement and proportional to the distance to point B, corresponding to the delamination process. The angular velocity $\widehat{\omega}$ is chosen such that the jump in velocity, between the fault GF and the BS, at point G is oriented according to case 2. The other points on the fault GF are in case 1, meaning that the fault sustains a mixed mode of shearing and opening. The upper bound to the tectonic force is

$$\begin{aligned} Q_u^{(3)} \cos(\phi_D) &= \rho g\big(S_{BS} \sin(\phi_D + \beta) + \widehat{\omega}[\cos\beta(L_{AB}S_{HW} - I_{1HW}) \\ &\quad + \sin\beta I_{2HW}]\big) + C_D L_{AG} \cos\phi_D + \widehat{\omega} T_D \frac{1}{2} L_{GB}^2 \\ &\quad + C_{GF} L_{GF}\left(\widehat{J}_G \cos\phi_{GF} + \frac{1}{2} L_{GF} \widehat{\omega} \operatorname{cotan}(\phi_{GF})\right), \\ \text{with} \quad 6I_{1HW} &= \left(L_{AB}^3 - [L_{AB} - x_{2F} \operatorname{cotan}(\alpha+\beta)]^3\right) \tan(\alpha+\beta) \\ &\quad - \left(L_{AG}^3 - [L_{AG} - x_{2F} \operatorname{cotan}(\theta_{GF})]^3\right) \tan\theta_{GF}, \\ \text{and} \quad 6I_{2HW} &= x_{2F}^3[\operatorname{cotan}(\alpha+\beta) - \operatorname{cotan}(\theta_{GF})], \end{aligned} \tag{31}$$

in which x_{2F}, I_{1HW}, and I_{2HW} are the second coordinate of point F and the moments of the HW with respect to the first and second axis, respectively. The norm of the velocity jump at G and the angular velocity are related by

$$\frac{\widehat{\omega} L_{GB}}{\sin(\phi_D + \phi_{GF} + \theta_{GF})} = \frac{1}{\cos(\theta_{GF} + \phi_{GF})} = \frac{\widehat{J}_G}{\cos\phi_D}, \tag{32}$$

in which $\widehat{J}_G$ is the norm of the velocity jump at point G and a normalization with respect to the BS velocity has been conducted. This upper bound is a function of the length L_{AB} and the angle θ_{GF}. These two parameters are varied making use of the same spatial discretization of the topography and the decollement as for CM 2. There is an optimum set of points (G,F) for which $Q_u^{(3)}$ in Eq. (31) is minimum and this value is denoted $Q_{lu}^{(3)}$.

This CM has the merits of its geometrical simplicity with the drawback that the fault *GF* is in a slip mode only at point *G* and is in a mixed mode of slip and opening along *GF* by construction. This mode of deformation is certainly not very realistic and the next CM is hoped to improve on this deficiency.

3.3 Collapse Mechanism 4: Delamination With a Curved Fault

CM 4, the second mechanism aimed to capture delamination, proposes a curved fault between the BS and the HW. The fault is tangent to the decollement at point G and its curvature is such that Coulomb friction is activated all along (Fig. 7D). The HW is rotating around point *B* such that the section *GB* of the decollement is delaminated, as for CM 3. This fault geometry is computed with a second-order asymptotic development in the arc length *s*, parameter of the fault from G to *F*. The application of the kinematic approach of LA provides the following upper bound

$$
\begin{aligned}
Q_u^{(4)} \cos(\phi_D) = {} & \rho g(S_{BS} \sin(\phi_D + \beta) + \widehat{\omega}[\cos\beta(L_{AB}S_{HW} - I_{1HW}) \\
& + \sin\beta I_{2HW}]) + C_D L_{AG} \cos\phi_D + \widehat{\omega} T_D \frac{1}{2} L_{GB}^2 \\
& + C_{GF} \cos\phi_{GF} \int_0^{s_F} \widehat{J}_{GF}(s')ds', \\
I_{1HW} = {} & \int_{HW} x_1 dS, \quad I_{2HW} = \int_{HW} x_2 dS, \\
\widehat{\omega} L_{GB} = {} & \frac{\sin(\phi_D + \phi_{GF})}{\cos\phi_{GF}},
\end{aligned}
\tag{33}
$$

in which $\widehat{J}_{GF}(s)$ and s_F are the norm of the velocity jump over *GF*, function of the arc length, and the arc length measured at point *F*, respectively. The moments I_{1HW}, I_{2HW}, the surface S_{HW}, and all quadratures are estimated by numerical means. The angle $\theta_{GF}(s)$, defining the orientation of the tangent to the fault at any s (see illustration in Fig. 7D) is obtained from the condition that the velocity jump along the fault is always in case 2

$$\tan(\theta_{GF}(s) + \phi_{GF}) = -\frac{\widehat{J}_{GF2}}{\widehat{J}_{GF1}}$$

$$\text{with} \quad \underline{\widehat{J}}_{GF}(\underline{x}) = -(\cos\phi_D - \widehat{\omega}x_2)\underline{e}_1 + [\widehat{\omega}(L_{AB} - x_1) - \sin\phi_D]\underline{e}_2, \tag{34}$$

in which (x_1,x_2) are the coordinates of the point on the fault GF at the arc length s. Note that the right-hand side of Eq. (34) consists of the ratio between the two components of the velocity jump along GF.

The minimization of the upper bound in Eq. (33) is conducted by computing for every point G on the descritized decollement, the fault geometry, and the associated upper bound. The least is denoted by$Q_{lu}^{(4)}$.

3.4 Collapse Mechanisms 5 and 6: Vergence Controlled by Compaction During Thrusting

The last two CMs are proposed to break the symmetric role of the two faults GE and GF in the CM 2. This symmetry breaking is done by assuming that one of the discontinuity is a ramp and deforming by Coulomb shear (case 2) whereas the material crossing the other discontinuity is in shear-enhanced compaction (case 4). The ramp is shearing repeatedly the same material and compaction cannot be the relevant deformation mechanism. To the contrary, the conjugate plane is crossed by new material constantly and this material could compact during its shearing.

These CMs 5 and 6 are explored assuming that the compaction strength P* in the bulk is not homogeneous but sustains a gradient typical of the compaction in a sedimentary basin. The necessity for this proposition is documented in the next section. This gradient is assumed to be a constant vector normal to the topography $\underline{\nabla P^*} = G^*\underline{N}$ where $\underline{N}$ is the normal to the topography pointing externally and G* the gradient intensity. The compaction strength at any point $\underline{x}$ within the wedge is then $P^{*0} + \underline{\nabla P^*} \cdot (\underline{x}_C - \underline{x})$ taking the point C as a reference point on the topography and P***0** standing for the surface value of the compaction strength. The application of the kinematic approach of LA provides the two upper bounds

$$
\begin{aligned}
Q_u^{(5)} \cos(\phi_D) =& \rho g\left(S_{BS} \sin(\phi_D+\beta) + \widehat{U}_{HW} S_{HW} \sin\left(\theta_{GE} - \phi_{GE}^* + \beta\right)\right) \\
&+ L_{AG} C_D \cos\phi_D + C_{GF} L_{GF} \widehat{J}_{GF} \cos\phi_{GF} \\
&+ \widehat{U}_{HW} \sin\phi_{GE}^*\left(L_{GE}\left[P_{GE}^{*0} + \underline{\nabla P}_{GE}^* \cdot (\underline{x}_C - \underline{x}_G)\right]\right. \\
&\left. -\frac{1}{2} L_{GE}^2 \underline{\nabla P}_{GE}^* \cdot \underline{t}_{GE}\right), \\
Q_u^{(6)} \cos(\phi_D) =& \rho g\left(S_{BS} \sin(\phi_D+\beta) + \widehat{U}_{HW} S_{HW} \sin(\theta_{GE} + \phi_{GE} + \beta)\right) \\
&+ L_{AG} C_D \cos\phi_D + C_{GE} L_{GE} \widehat{U}_{HW} \cos\phi_{GF} \\
&+ \widehat{J}_{GF} \sin\phi_{GF}^*\left(L_{GF}\left[P_{GF}^{*0} + \underline{\nabla P}_{GF}^* \cdot (\underline{x}_C - \underline{x}_G)\right]\right. \\
&\left. -\frac{1}{2} L_{GF}^2 \underline{\nabla P}_{GE}^* \cdot \underline{t}_{GF}\right),
\end{aligned}
\tag{35}
$$

where $\underline{t}_{GF}$ and $\underline{t}_{GE}$ are the unit vectors tangent to the fault indicated in subscript and oriented from G toward the free surface. The velocity norms in Eq. (35) are normalized with respect to the BS velocity according to

$$
\begin{aligned}
\text{CM 5}: \quad & \frac{\widehat{J}_{GF}}{\sin\left(\theta_{GE} - \phi_{GE}^* - \phi_D\right)} = \frac{1}{\sin\left(\theta_{GF} + \phi_{GF} + \theta_{GE} - \phi_{GE}^*\right)} \\
& = \frac{\widehat{U}_{HW}}{\sin(\theta_{GF} + \phi_{GF} + \phi_D)},
\end{aligned}
\tag{36}
$$

$$
\begin{aligned}
\text{CM 6}: \quad & \frac{\widehat{J}_{GF}}{\sin(\theta_{GE} + \phi_{GE} - \phi_D)} = \frac{1}{\sin\left(\theta_{GF} - \phi_{GF}^* + \theta_{GE} + \phi_{GE}\right)} \\
& = \frac{\widehat{U}_{HW}}{\sin\left(\theta_{GF} - \phi_{GF}^* + \phi_D\right)}.
\end{aligned}
$$

The minimization of Eqs. (35) and (36) is done as for mechanism 2: the load is computed for every set of point (E,F,G) and the selected set is the one leading to the minimum bound.

3.5 Triangular Wedge: Predicting the Dominant Mechanism

Before starting the discussion on the comparison between these six CMs, the following dimensional analysis is proposed. Any physical quantity such as A has a reference value A_R and its dimensional expression is $\tilde{A} = A/A_R$.

The reference stress σ_R is taken equal to the bulk cohesion C set to 10 MPa. The material weight ρg is 2510^4 Pa/m so that the ratio $\rho g/\sigma_R$ provides the characteristic length $L_R = 400$ m. Material and geometrical parameters are summarized in Table 1.

3.5.1 Collapse Mechanisms 1 and 2 for Comparison With the Critical Coulomb Wedge Theory

The first objective is to compare the predictions of the CM 1 and 2 with the critical taper theory. The ratio of least upper bounds $Q_{lu}^{(2)}/Q_{lu}^{(1)}$ is presented as a function of the topographic slope α in Fig. 8A. The distance from the back wall to the root of the thrust fold L_{AG} is found also as a function of α in Fig. 8B. The dashed curves correspond to the cohesionless wedge, and these results are analyzed first. The bound ratio is increasing approximately linearly for small values of the topographic angle, and the distance L_{AG} is small compared to L_{AB} signaling that the optimum thrust folds for each α is at the back of the wedge, as expected for subcritical wedges ($\alpha < \alpha_c$). The force ratio is equal to one exactly for the topographic slope $\alpha_c = 2.106°$ predicted by the critical taper theory. The length L_{AG} for α_c is indeterminate and jumps from a small value to a value close to L_{AB}. The position of the thrust fold can be anywhere within the wedge. For larger α, the force ratio is one meaning that the whole wedge is gliding on the decollement and the deformation is indeed found at the front, up to numerical discretization error. The topographic conditions are thus

Table 1 Geometrical and material properties for the triangular wedge and the six collapse mechanisms presented in Fig. 7 unless they are varied (var) in the analysis

Symbol	Definition	Value	Unit
β	Decollement dip	2	deg
ρg	Material unit weight	2500	Pa/m
ϕ	Bulk friction angle	30	deg
ϕ^*	Bulk compaction angle	40	deg
C	Bulk cohesion	10	MPa
P^*0	Bulk compaction strength at surface	var	MPa
G^*	Bulk compaction strength gradient	var	MPa
ϕ_D	Decollement friction angle	10	deg
C_D	Decollement cohesion	var	MPa
T_D	Decollement tensile strength	0	MPa

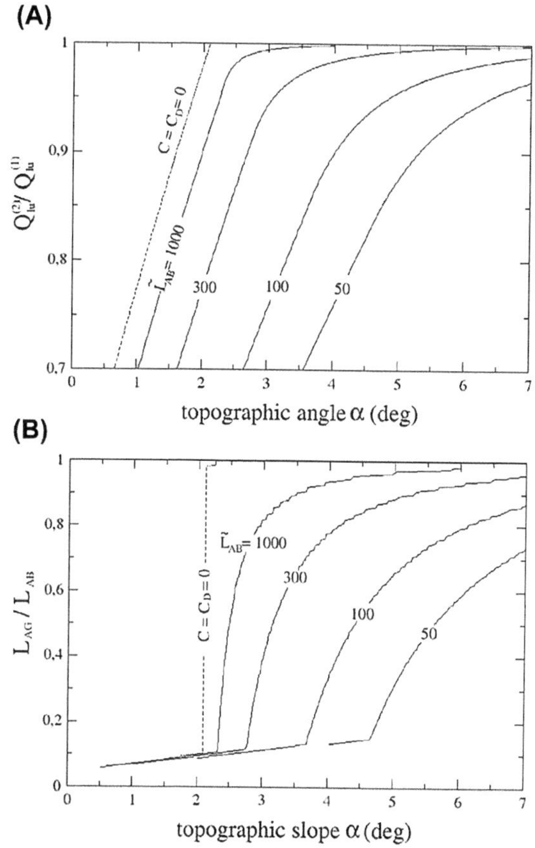

Figure 8 Super- and subcritical conditions defined by collapse mechanisms(CMs) 1 and 2 for a cohesive and frictional wedge of finite extent. The influence of the cohesion is felt by varying the decollement length L_{AB}. The *dashed curves* correspond to a cohesionless wedge. The ratio of the least upper bounds (CMs 2 and 1) is presented in (A) and the extent of the activated decollement in (B).

supercritical ($\alpha > \alpha_c$). CM 1 and 2 are thus in agreement with the classical CCW theory, exact only for cohesionless wedges of infinite extent.

The criticality conditions accounting for cohesion are presented in the same Fig. 8 with $\tilde{C}_D = \tilde{C} = 1$. The wedge length L_{AB} is varied between 50 and 1000 times the reference length L_R of the problem. Cohesion should only be important for small value of $\tilde{L}_{AB}$. The first observation is that the force ratio is not anymore a bilinear function of the topographic slope. There is a gradual transition from the approximately linear relation for small α to the horizontal asymptote at one for large values of α. This transition is smoother for the smaller values of $\tilde{L}_{AB}$. In the limit of $\tilde{L}_{AB}$ tending to infinity, the solution converges toward the results obtained in the absence of cohesion. The smooth transition from sub- to supercritical conditions is due to the progressive migration of the dominant thrust fold from the rear to the front of the wedge with increasing values of α, Fig. 8B. Note that this transition is not activated at α_c but at α_c plus a positive difference which is increasing for the smaller values of $\tilde{L}_{AB}$.

The cohesion of the decollement is now varied with respect to the bulk material cohesion for the two values of the length $\tilde{L}_{AB}$ of 50 and 300. The results are presented in Fig. 9. Consider first the results for $\tilde{L}_{AB} = 50$, the dotted–dashed curves presented for $\tilde{C}_D = 1, 0.5$, and 0. Reducing the cohesion by half has shifted the curve of the bound ratio toward the

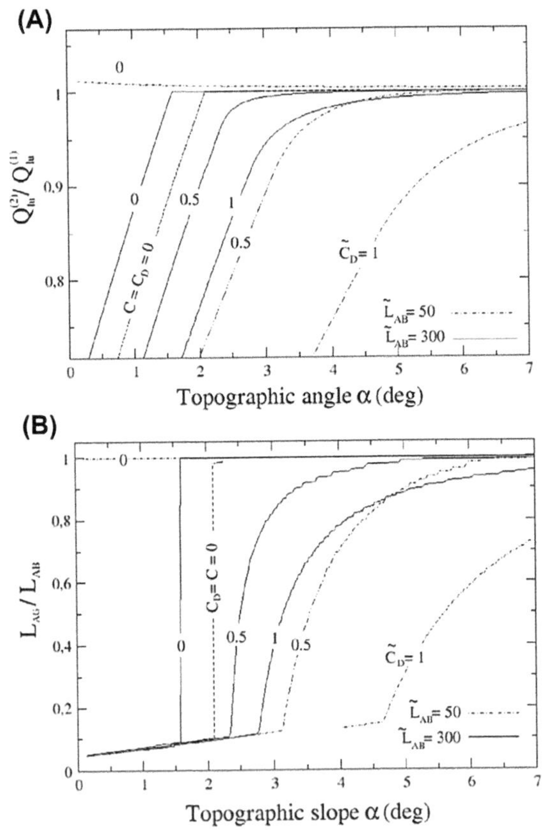

Figure 9 Influence of the decollement cohesion on the cohesive wedge criticality conditions for two lengths $\tilde{L}_{AB} = 50$ and 300 in *dotted–dashed* and *solid curves*, respectively. The results for the cohesionless wedges are the *dashed curves*.

critical taper solutions. There is still a delay in the beginning of the smooth transition in the position of the root G from the rear to the front of the wedge but this delay is reduced by half, Fig. 9B. The next value of the decollement is $C_D = 0$ and in that instance the least upper bound for CM 2 is always greater than one. This limit of one is exceeded because of the numerical discretization but the important result is that the cohesive wedge on the cohesionless decollement with $\tilde{L}_{AB} = 50$ is always supercritical. This conclusion is confirmed by the length of the activated decollement, which is found to be equal to the total length of the wedge, Fig. 9B. Complementary results are presented for the length $\tilde{L}_{AB} = 300$ solid curves. The new information is that the bound ratios for $\tilde{C}_D = 0.5$ and 0 are around the critical taper solutions. Note that the criticality conditions for the cohesionless decollement occur for an angle approximately 0.5° less than for the critical taper theory.

3.5.2 Delamination: Mechanisms 3 and 4

The least upper bounds for CM 3 and CM 4, two mechanisms that introduce a delamination in the frontal part of decollement with either a straight or a curved hinterland ramp, are compared with CM 2 (frictional foreland thrust) in Fig. 10. The forces are normalized with respect to the upper bound for CM 1. It should be noted that the delamination mechanisms can only compete with CM 2 for wedges sufficiently small. It is for this reason that results are presented for $\tilde{L}_{AB} = 10$ in Fig. 10.

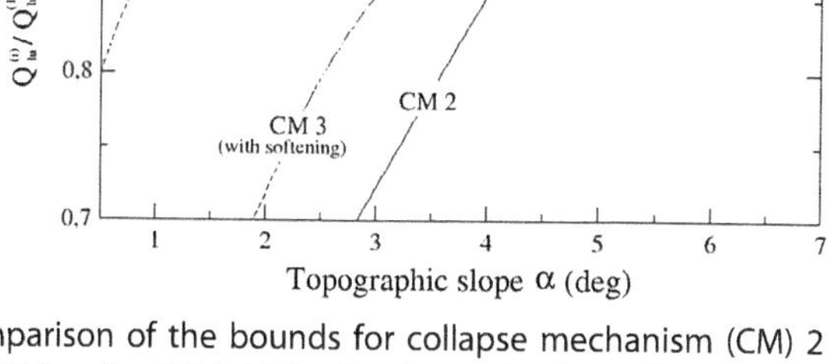

Figure 10 Comparison of the bounds for collapse mechanism (CM) 2 (frictional foreland thrust), *solid line*; for CM 3 (delamination with a hinterland straight ramp), *dashed curves*; and CM 4 (delamination with a curved hinterland ramp). The wedge length is $\tilde{L}_{AB} = 10$ and the softening for the delamination mechanisms is such that the fault *GF* has the properties of the decollement instead of the bulk material.

The first set of results is obtained assuming the straight or curvy fault GF has the properties of the bulk material. The ratio of bounds for CM 3 and 4 is always larger than the ratio for the classical thrusting CM 2. They cannot be activated. Moreover, CM 3 that is supposed to improve on CM 2 was always supercritical, positioned to the very front of the triangular wedge and with a bound ratio equal to one. A second set of predictions is based on some softening on the fault GF. The fault is assigned, for sake of simplicity, the same properties as the decollement. The results are more satisfactory in the sense that CM 3 dominates CM 2 for α greater than approximately 5°. The root G is at the dimensionless distance from the BS of 8 and 5, for CM 3 and CM 2, respectively, for this specific angle. For this range of topographic angle, CM 4 is supercritical and found at the very tip of the wedge.

3.5.3 Compaction and Hinterland Thrusting: Collapse Mechanisms 5 and 6

The comparison between CM 5, 6, and 2 is proposed to question the vergence of the thrusting. CM 2 considers frictional response on its axial surface and ramp, the two having the same role. CM 5 and CM 6 introduce shear-enhanced compaction on the axial surface so that thrusting is hinterland and foreland, respectively. Results are presented in Fig. 11 for $\tilde{L}_{AB} = 100$. Two sets of results are presented, the first in the absence of compaction gradient $\tilde{G}^* = 0$ and the second with a gradient of 3. The results for CM 2 correspond to the solid curves.

The first set of results is interesting since it shows that the two thrusting mechanisms with compaction dominate with a preference for CM 5 proposed for a hinterland vergence. Nevertheless, the position of the root of this mechanism is at the back of the wedge for all topographic angles. It is thus not associated with a frontal deformation. Introducing a gradient $\tilde{G}^*$ has the definite advantage to penalize the deformation at the back. There is indeed now a criticality transition, which is rather close to the one recorded for the classical CM 2. The difference in topographic angle at this criticality transition between CM 2 and CM 5 is not exceeding 1°. The bound ratio is still in favor of the two compacting mechanisms with a slight advantage for CM 5, which promotes hinterland vergence.

3.5.4 Discussion

These comparisons have brought four important conclusions, which are now presented. First, the kinematic approach of LA reproduces exactly

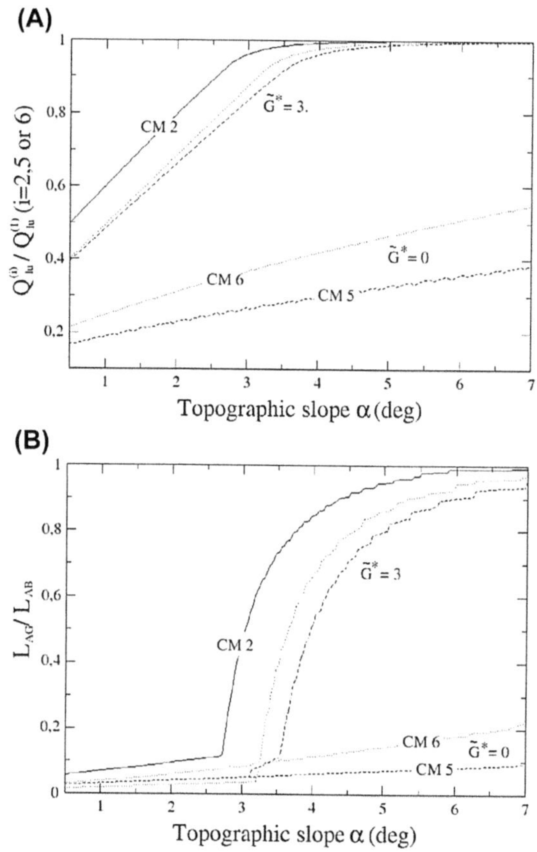

Figure 11 Comparison of collapse mechanisms 5, 6, and 2, to question the thrusting vergence, in terms of the bound ratio (A) and the position of the root G on the decollement (B). Results obtained for $\tilde{L}_{AB} = 100$.

the criticality conditions of the CCW theory, as presented many times since the early work of Cubas et al. (2008). If the topographic slope is increased beyond α_c, the position of the root of the dominant frictional thrust (CM 2) jumps from the back to the very front of the wedge.

Second, cohesive wedges can indeed be treated exactly with the kinematic approach of LA. Cohesions of the bulk and the decollement ($C = C_D$) are increasing the transition from subcritical to supercritical conditions by several degrees, the effect being important for wedge lengths 50 times the characteristic length $C/\rho g$. The transition is not abrupt but smooth as marked by the root of the critical thrust on the decollement, which shifts gradually from the back to the front as α is increased. Setting the decollement cohesion to zero, while the bulk is still cohesive, brings back the sudden jump typical of the criticality transition. This is true for large wedges (300 times the characteristic length) but not for smaller wedges (50 times only) since then the wedge is always supercritical.

The third conclusion here is that the delamination of the decollement does not suffice to promote the hinterland thrusting either on a flat or a curvy ramp. Softening and a large topographic angle are necessary. This softening signals the importance of the heritage: the presence of a fault having been activated prior to this setting.

The fourth conclusion is that a difference in deformation mechanism along the ramp and the axial surface could explain the hinterland vergence of some thrusting. Shear-enhanced and frictional response would characterize the axial surface while the ramp would be frictional only.

4. AN ACCRETING WEDGE AND TWO DECOLLEMENTS

The last feature of the triangular zone, the change in decollement depth close to the accreting front, is studied in this section. Only frictional properties are considered (case 2 of the support function for a Coulomb criterion).

The front of our wedge is presented in Fig. 12A where the accreting layer has the thickness H with respect to the lower decollement (properties ϕ_D, C_D). The length of the wedge from the back wall to the point I, break point between the topographic slope α and the accreting front, is denoted L_I. The second decollement (dashed line) is parallel to the first decollement at the distance h from the accretion topography and has the properties (ϕ_D', C_D'). Two CMs of the type CM 2, defined in the previous section, could be proposed for the activation of either the lower or the upper decollement. They both terminate at the front by thrusting to the free surface: points F_l, G_l, and E_l and points F_u, G_u, and E_u in Fig. 12A, the subscript l and u referring to the lower and upper decollement, respectively. The upper mechanism is nevertheless not possible since the whole wedge is compressed and the lower decollement has to be activated at least over a minimum length to ensure the transfer of activity to the upper decollement. A new CM is thus required, CM 7, which is presented in Fig. 13B for the particular case where the transfer occurs close to the wedge front. The fault GF is again an axial surface separating the BS from the lower HW region (HW_l). The ramp of this lower thrusting is however not cutting through

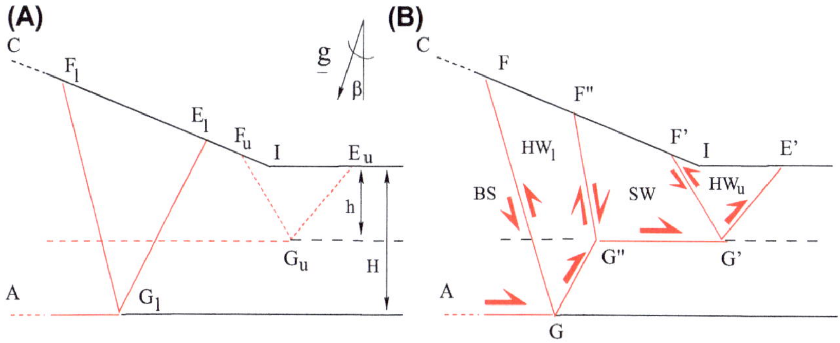

Figure 12 Prototype to capture the transition in decollement depth at the front of the accreting wedge. Two decollements are in competition with a thrusting to the free surface according to collapse mechanism (CM) 2 (A). The new mechanism, CM 7, includes a transition from the lower to the upper decollement and a thrusting to the free surface at the front (B).

the entire structure but stops as it reaches the upper decollement at point G''. The sliding wall (SW) above the segment $G''G'$ is gliding rigidly over the upper decollement. It is limited at its rear by the discontinuity $G''F''$, which sustains a normal sense of shear and which is also an axial surface. The point G' is the root of the two discontinuities $G'F'$ and $G'E'$ bounding the upper HW (HW_u) that defines the most frontal part of the thrusting mechanism. The former is an axial surface with a reverse sense of shear and the latter acts as a ramp. The application of the kinematic approach is rather systematic and provides the following expression for the upper bound to the force necessary to activate this new CM:

$$
\begin{aligned}
Q_u^{(7)}\cos(\phi_D) = \rho g\Big(& S_{BS}\sin(\phi_D+\beta) + \widehat{U}_{HW}S_{HWl}\sin(\theta_{GG''}+\phi_{GE}+\beta) \\
& + \widehat{U}_{SW}S_{SW}\sin\left(\phi_D'+\beta\right) + \widehat{U}_{HWu}S_{HWu}\sin\big(\theta_{G'E'} \\
& + \phi_{GE}+\beta\big)\Big) + L_{AG}C_D\cos\phi_D + L_{G''G'}C_D'\cos\phi_D' \\
& + C_{GE}\cos\phi_{GE}\left(L_{GG''}\widehat{U}_{HWl} + L_{G'E'}\widehat{U}_{HWu}\right) \\
& + C_{GF}\cos\phi_{GF}\left(L_{GF}\widehat{J}_{GF} + L_{G''F''}\widehat{J}_{G''F''} + L_{G'F'}\widehat{J}_{G'F'}\right).
\end{aligned}
\quad (37)
$$

The two first lines in the right-hand side of Eq. (37) correspond to the work of gravity, the third line to the resisting power of the two activated sections of the decollements, and the last two lines to the resisting power of the ramps and the axial surfaces. Note that all axial surfaces and ramps

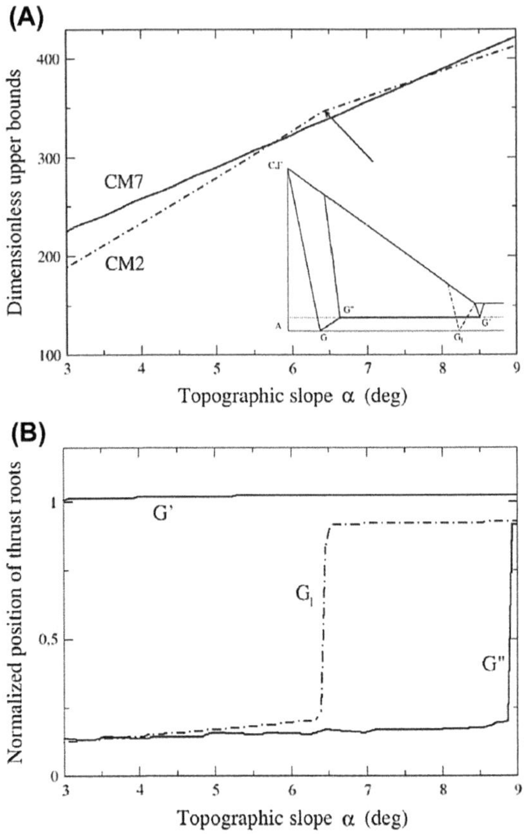

Figure 13 The bound for the change in decollement [collapse mechanism (CM) 7] is compared to the bound for classical thrusting on the lower decollement (CM 2), the two forces being function of the topographic slope (A). Note the small interval for dominance of CM 7. Note also in the inset that the transition occurs at the rear of the wedge in that interval. The position of the three points G, G', and G'' are presented in (B) as a function of α to discuss the depth change.

are assigned the frictional properties (ϕ_{GF},C_{GF}) and (ϕ_{GE},C_{GE}), respectively, corresponding to the properties introduced for CM 2. The velocity norms in Eq. (37) are defined by

$$\frac{\widehat{J}_{GF}}{\sin(\theta_{GG''} + \phi_{GE} - \phi_D)} = \frac{1}{\sin(\theta_{GF} + \phi_{GF} + \theta_{GG''} + \phi_{GE})} = \frac{\widehat{U}_{HWl}}{\sin(\theta_{GF} + \phi_{GF} + \phi_D)}, \tag{38}$$

$$\frac{\widehat{J}_{G''F''}}{\sin\left(\theta_{GG''}+\phi_{GE}-\phi_D'\right)}=\frac{\widehat{U}_{SW}}{\sin(\theta_{G''F''}+\phi_{GF}+\theta_{GG''}+\phi_{GE})}$$
$$=\frac{\widehat{U}_{HWl}}{\sin\left(\theta_{G''F''}-\phi_{GF}+\phi_D'\right)},$$

$$\frac{\widehat{J}_{G'F'}}{\sin(\theta_{G'E'}+\phi_{GE}-\phi_D)}=\frac{\widehat{U}_{SW}}{\sin(\theta_{G'F'}+\phi_{GF}+\theta_{G'E'}+\phi_{GE})}$$
$$=\frac{\widehat{U}_{HWu}}{\sin\left(\theta_{G'F'}+\phi_{G'F'}+\phi_D'\right)},$$

with the same notation used before, any angle θ_{XY} defines the slope of the generic discontinuity XY. Virtual velocity is normalized with respect to the BS velocity.

The technical difficulty with this mechanism concerns the minimization of the upper bound in Eq. (37). There are three axial surface dips, two ramp dips, and the two lengths of the activated sections of the decollement, which are unknown. These seven unknowns are equivalently replaced by the unknown positions of the seven points F, F'', F', E', G, G'', and G'. The computation time required for optimization is of the order of M^N where M is the number of points on the discretized surfaces and $N=7$ the number of unknown points. Such a calculation could be done faster on a graphical processing unit architecture but a simplified approach is proposed in this contribution. The selection of the frontal part of the thrusting mechanism (points E', G', and F') is independent of the four others and determined directly from the optimization of CM 2 proposed in Fig. 12A for the upper decollement. The computation cost is then reduced to M^4+M^3 and is not an issue anymore. This proposition is not steered only by these computational difficulties but also by the observation that the upper thrusting part of CM 7 is identical to CM 2 with the only difference that the BS velocity of CM 2 is now the velocity of the SW. It is thus legitimate to assume that the optimum positions of points E', G', and F' could be obtained by isolating the frontal part of CM 7. The positions of E', G', and F' are thus taken as those of E_u, G_u, and F_u defined in Fig. 12A.

The first results constitute a comparison between CM 7 and CM 2 for the geometrical and material parameters provided in Table 2. The upper bounds for the two mechanisms are provided in Fig. 13A as a function of the topographic slope α of the triangular part of the wedge. The dotted–dashed curve for CM 2 is below the CM 7 curve for most of the range of

Table 2 Geometrical and material properties to study the change in decollement depth for the wedge with an accreting front

Symbol	Definition	value
$\tilde{L}_I$	Length of triangular region	100
$\tilde{H}$	Accreting layer thickness	2.5
$\tilde{h}$	Depth of upper decollement	1.25
ϕ_D	Lower decollement friction angle	20°
$\tilde{C}_D$	Lower decollement cohesion	0.5
$\phi_D^{'}$	Upper decollement friction angle	10°
$\tilde{C}_D^{'}$	Upper decollement cohesion	0

The missing information is found in Table 1. The reference stress and length are still $\sigma_R = 10$ MPa and $L_R = 400$ m.

α considered. There is, however, a window in the range [6°;7.5°] for which the change in decollement can indeed occur. The inset within this figure shows the position of the two CMs for the slope of 6.5° within this small interval. The transition from the lower to the upper decollement occurs at the rear of the wedge and CM 7 makes use mostly of the upper decollement up to the thrust at the very front of the wedge. The ramp of CM 2 and the point $F^{'}$, tip of an axial surface of CM 7, are exactly at the break slope point I.

These results could be explained with the CCW theory noting that $\alpha_c = 6.36°$ and $\alpha_c^{'} = 2.10°$ for the lower and the upper decollement, respectively. For that purpose, the x_1-coordinate of points G', G'', and G_l, normalized by the length L_I, have been plotted in Fig. 13B as a function of the slope α. The coordinate G' of the upper thrusting part of CM 7 is always greater than one since it is found on the accreting layer. This upper thrust is always supercritical since the values of α are exceeding $\alpha_c^{'}$. The choice of a zero cohesion is certainly responsible for the activation of the upper decollement beyond the point I. Point G_l is close to the back of the wedge for $\alpha \leq \alpha_c$ and migrates abruptly to the front as soon as α exceeds α_c, as expected. It is around this change in position that CM 7 appears to be dominant. However, the position of point G'' signaling the transition to the upper decollement is at the back of the structure. One has to consider larger topographic slope around $\alpha = 9°$ for this point G'' to be found at the front of the accreting wedge. Note, however, that the length $G''G'$ of the activated upper decollement section is then small. It is for that reason that the change in decollement is not so beneficial and that the dominant mechanisms remain CM 2 for these large topographic slopes.

Our prototype is certainly too simple to explain the frontal position of CM 7 and its dominance over CM 2. Several reasons could be evoked to explain this result including a gradient in frictional properties. One could imagine a negative gradient in the frictional properties toward the front, which would penalize the transition from the lower to the upper decollement at the rear of the wedge. One could also imagine that the transition ramp GG'' is not a new fault but an inherited structure reactivated in compression. Another explanation is that the shape of the wedge at the front is more complex than idealized in our prototype on a length scale corresponding to the activated section $G''G'$ of the upper decollement.

To investigate this possibility, a trilinear wedge has been considered, as seen in Fig. 14. The slope of the segment KI is kept constant at 5° to make sure that the upper thrusting part of CM 7 is at the very front. The slope of the rear part of the topography CK is changed over a small interval around the 9°, which was found necessary to obtain a point G'' at the front in our first set of results. The shape of CM 7 and CM 2 are presented for $\alpha = 8.9°$ and $\alpha = 10°$ in solid and long dashed lines, respectively. For the slope of 8.9°, points G and G'' are at the coordinates 62 and 68 on the lower and the upper decollement, respectively. The section of the upper decollement which is activated, $G''G'$ is below the KI segment of the triangular region. The V-shaped thrust of CM 2 is also presented on that figure for the same topography angle. The point G_l is ahead of G at the coordinate 67, so CM 2 requires a longer activation of the lower decollement than CM 7. The results for $\alpha = 10°$ differ by the relative position of these two points; G is now very close to G_l and has shifted by a distance of 6 despite the small angular increase of 1.1°. Such a large variation is typical at the criticality transition.

These preliminary results should be completed by a thorough parametric study but two results seem already to emerge. First, the change in decollement could dominate close to the criticality conditions based on the lower decollement. The transition from the lower to the upper decollement occurs at the back of the wedge at these critical conditions and to the front of the wedge for topographic slopes much larger (close to $1.5\alpha_c$ for our parameters). Second, a curvature or a change in the topographic slope in the frontal region can provide the characteristic length controlling the extent of the activated section of the upper decollement.

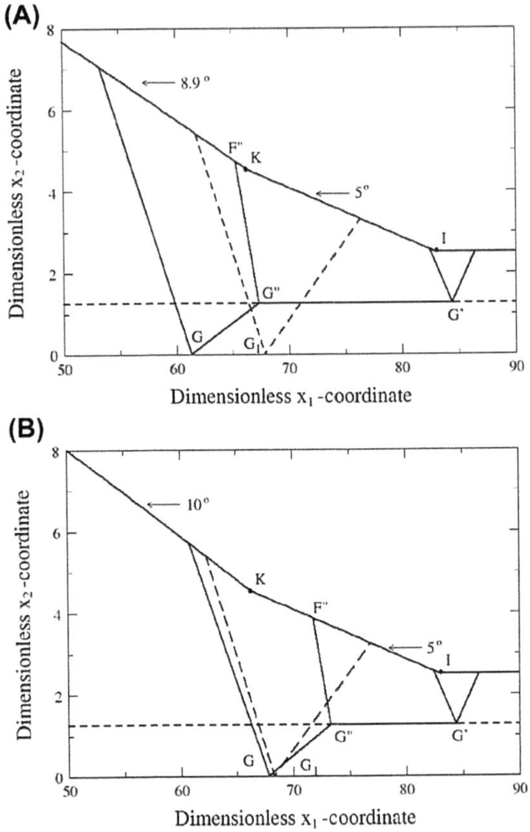

Figure 14 A trilinear topography is proposed to capture the change in depth of the active decollement in the accreting wedge, based on collapse mechanism 7. The rear section dips at 8.9° and 10° in (A) and (B).

5. SEQUENTIAL LIMIT ANALYSIS

The main objective of this contribution has been so far to illustrate the potential of the kinematic approach of LA in selecting the CM, which characterizes the dominant mode of deformation. This section complements this discussion by reviewing how to construct the evolution of the geological structure based on the dominant CM.

The simple example of two competitive folds, presented in Fig. 15, is proposed for that purpose. Let us consider a layer resting on a weak decollement and assume that a CM typical of a thrust at its onset has been detected, Fig. 15A. The fold in Fig. 15B is changing its geometry

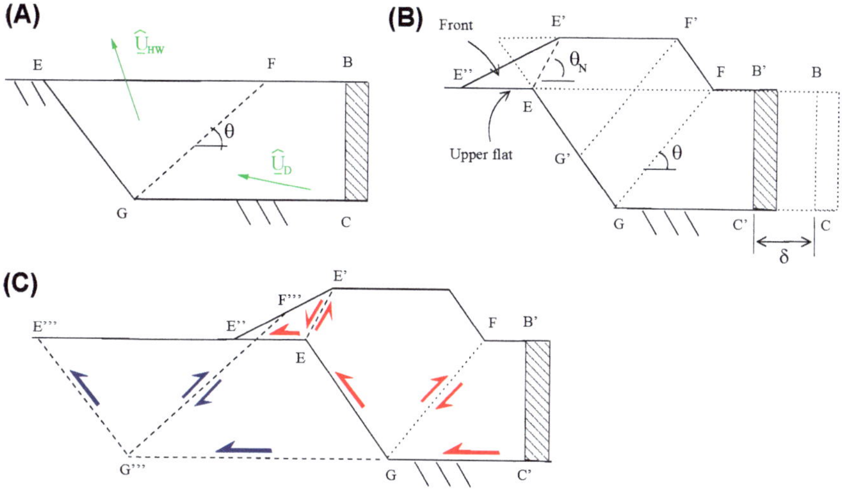

Figure 15 The sequential limit analysis with the simple example of a sequence of two fault-bent folds. The collapse mechanism determining the first thrust, in the first step (A). The fold construction according to simple geometrical rules, in the second step (B). At each increment of shortening, the bound to the tectonic force for the current [red arrows] and a new thrust [blue arrows] are compared (C).

with shortening δ according to simple geometrical rules. The material in the BS (GCBF) is moving parallel to the decollement accommodating this increment of shortening. Material that goes through the axial surface GF takes a displacement direction parallel to the ramp GE with an increment of size proportional to δ to ensure mass conservation. Material already present in the HW is translated by the same displacement explaining why the initial axial surface GF is now at G$'$F$'$. The material displacement within the HW has for consequence that its tip (previously at point E) is now up above the topography (dotted lines). This anomaly is fixed by proposing to project the upheaved material back on the upper flat as if it had been sheared in a normal sense through the new axial surface EE'. The dip of EE', θ_N could be chosen by optimization of the upper bound if a friction coefficient on the upper flat is proposed. It could also be chosen such that EE$'$ bisects the angle $\widehat{G'EE''}$. The material in the front is then sliding on the upper flat during further shortening.

The upper bound which was computed at the onset of thrusting can still be evaluated during the fold development and the sources of dissipation are presented in Fig. 15C. These sources consist of the axial surface GF, the ramp GE, the upper and lower decollement EE'' and GC', and the upper axial surface EE'. The associated velocity field is readily constructed following the same approach as in the previous sections. The question of the life span of this fold is resolved by searching for a new fold providing an upper bound less than the one just computed. A candidate is proposed in Fig. 15C. It makes use of a longer segment of the decollement $G'''C'$ and thrusts to the top surface, thanks to the ramp $G'''E'''$ and the axial surface $G'''F'''$ intersecting the frontal region of the first thrust. It is found by varying the positions of points E''', F''', and G' on the discretized decollement and topography for sake of optimization.

It is this incremental scheme composed of the search for the optimum CM and the folding development based on simple geometrical rules that is called SLA.

The first application of the SLA for fold-and-thrust belts was proposed by Cubas et al. (2008), for a wedge with an accreting front, Fig. 16A. The two horizontal lines parallel to the weak decollement are deformation markers. Material parameters are selected such that the wedge is supercritical and the first CM is at the front, defining the emplacement of the first thrust T_1. The fold T_1 accommodates some shortenings and is in competition with any new thrust with a similar CM as for its onset. This competition is sensitive to the material properties of the dissipative sources. For example, the first thrust is abandoned for a new thrust positioned at its front, at the shortest distance allowed by the discretization of the decollement and of the topography, if the material properties remain constant during the shortening. The first fold has a finite amplitude only if some softening on the material properties are accounted for to counterbalance the increasing tectonic force due to the increasing mass present in the HW and on the frontal part. Cubas et al. (2008) chose to reduce the friction angle of the ramp. This softening is assumed to occur at the first increment of shortening beyond the onset explaining the three drops in the upper bound to the tectonic force presented in Fig. 16B. Reducing friction on the ramp facilitates slip during shortening, reducing the tectonic force despite the increasing mass in the HW. The thrust T_1 has thus a finite activation time span given by the normalized shortening $\tilde{\delta}_1$. It is for this shortening that the onset of thrust T_2 is detected at the front of T_1 with an axial surface outcropping at the tip of the frontal region of T_1. Fold T_2 develops leading to a complex geometry in the old T_1 material, which is now convected through its axial surface. The second fold develops up to the shortening $\tilde{\delta}_2$ at which T_3 becomes dominant and starts to accommodate the overall shortening.

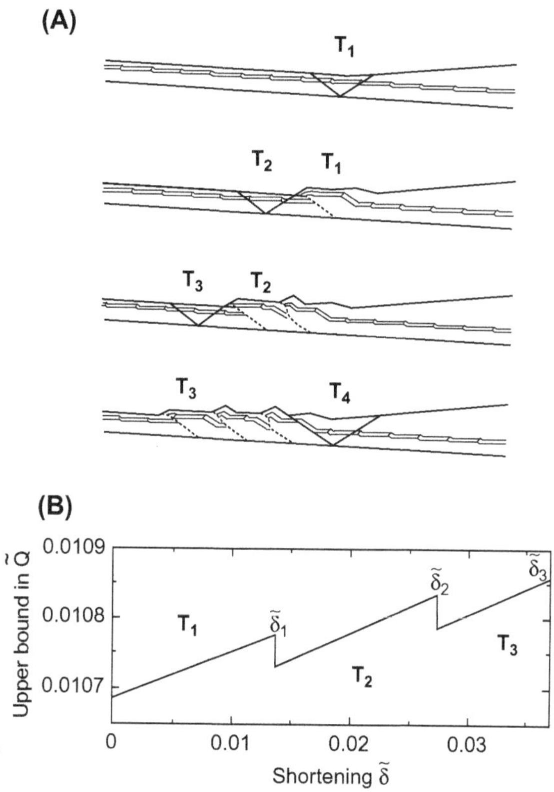

Figure 16 A forward sequence of three thrusts up to the onset of the first out-of-sequence thrust T_4 at the back (A). The two lines initially parallel to the decollement are markers. The least upper bound to the tectonic force during shortening, the three drops corresponding to the frictional softening considered on the ramp of the active fold (B). *From Cubas, N., Leroy, Y., & Maillot, B. (2008). Prediction of thrusting sequences in accretionary wedges.* Journal of Geophysical Research: Solid Earth, 113*(B12).*

This forward sequence ceases with the detection of T_4 in the basin formed at the back of T_1. There is a wealth of information within such sequences presented in Cubas et al. (2008), but the technical reason to end the simulation is due to the geometrical rules presented above. The frontal part of the thrust was designed to override a flat topography, the upper flat, and T_4 needs to overthrust a complex topography, which is certainly not flat. This technical difficulty was suppressed by Mary et al. (2013a) who proposed to deposit the frontal part on the irregular topography as if it was a granular material of negligible cohesion, as for sand in analog laboratory experiments. The slope of the material newly deposited on the topography is set to the repose angle (the friction angle) and the position of this slope is selected to impose mass conservation. With such minor modifications of the geometrical rules it is not four thrusts that can be captured but unlimited sequences of thrusting events.

Mary, Maillot, and Leroy (2013b) also proposed to include erosion laws following the ideas of Konstantinovskaya and Malavieille (2011). Erosion occurs periodically by removing material found above a line. This line is going through the same material point selected in the BS region where there is no deformation. The line is also defined by the slope α_e. Many simulations were conducted in trying to reproduce the deformation style reported by Konstantinovskaya and Malavieille (2011) and presented in Fig. 17A. The parameter variations is, of course, necessary in view of the uncertainty attached to the analog modeling and also since some of the data were not reported by Konstantinovskaya and Malavieille (2011) such as the residual friction ϕ_{Rf}. The decollement friction angle and the erosion slope were also varied and three simulation results are presented in Fig. 17B–D.

The principal observation is that the successive thrusting in the central region of the box has resulted in a large number of thrust sheets which were progressively rotated. Some are now vertical and their material initially in

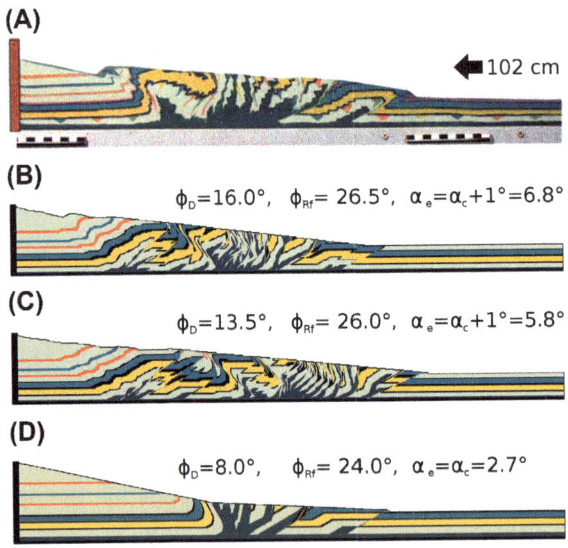

Figure 17 Comparison between the experimental result of Konstantinovskaya and Malavieille (2011) (A) and the numerical result of Mary et al. (2013b) for different values of the erosion slope α_e, residual sand friction ϕ_{Rf}, and decollement friction ϕ_D (B) and (C). *From Mary, B. C., Maillot, B., & Leroy, Y. M. (2013b). Predicting orogenic wedge styles as a function of analogue erosion law and material softening.* Geochemistry, Geophysics, Geosystems, 14*(10), 4523–4543.*

contact with the decollement is now piercing the topography due to the erosion. The thickness and the number of these thrust sheets as well as the presence of an isolated thrust at the front compared well with the analog experiments in the case of a small detachment friction and an erosion slope equal to the CCW slope ($\alpha_e = \alpha_c$), Fig. 17D. Increasing the basal friction, reducing the fault softening, and increasing the erosion slope above α_e result in a more complex deformation style with, in particular, the frontal fold being segmented into two thrusts in Fig. 17B and being less apparent in (C).

The striking information of this comparison is that minor changes in mechanical and erosive properties have a huge impact on the deformation style. It is only by conducting a large number of simulations in the spirit of inverse methods (Tarantola, 2005) with some observables as target (for example, presence of an isolated thrust at the front, number and thickness of thrust sheets reaching the top surface in the central region) that these parameter ranges could be determined (Cubas, Barnes, & Maillot, 2013).

In the meanwhile, there is another striking information revealed by the numerical simulations in Fig. 17. The layers initially flat and parallel to the decollement have been disrupted and intersected by the thousands of active axial surfaces and ramps. Keeping track of this mosaic of subdomains is a difficult task from the numerical point of view. Although it was done here for sake of comparison with the analog experiments, this numerical issue will become one of the two main technical locks if each layer would have specific material properties. The second technical lock is that such heterogeneity, which is a characteristic of the upper crustal materials, would require a CM with as many changes in ramp orientation as the number of crossed subdomains. The optimization would not be based on three degrees of freedom, as typically done here, but thousands, requiring a sophisticated optimization scheme.

6. CONCLUSION

The first of the two objectives of this contribution was to provide the fundamentals in mechanics to master the kinematic approach of LA. It is hoped that this text could be used by scientists interested to teach the subject at the graduate level. In the process, it was shown that LA does reproduce the classical results of the CCW theory using a simple CM corresponding to a V-shaped thrust with two straight discontinuities interpreted as a ramp and an axial surface, the two discontinuities rooting at the same point on the decollement. It was also shown how this LA can be used creatively to propose new CMs aimed at shedding light on the onset of geological structures. The second objective was to open up the application range of LA and to show that it could be combined with classical geometrical constructions of fault-related folds. The resulting scheme, called SLA, permits to follow the long-term evolution of an accretionary wedge or of a fold-and-thrust belt, while accounting for erosion and deposition.

To illustrate our claim, we have considered the open problem of the onset of triangular zones at the front of accreting wedges. Three technical questions were raised, first the importance of delamination (small tensile strength) with a straight or a curved ramp leading to a hinterland thrusting. The surprising result is that delamination as a deformation mechanism along the frontal part of the decollement, with either a straight or a listric ramp, does not help to overcome the dominance of the classical V-shaped thrust. The only case where delamination and a hinterland ramp can be dominant is if the latter is an inherited discontinuity with lower frictional properties as the surrounding bulk material.

The second technical question was how to break the symmetry between the two frictional discontinuities constituting the V-shaped thrust discussed earlier. Which discontinuity is a ramp or an axial surface? Is the thrusting hinterland or foreland? The kinematic approach of LA cannot differentiate between the two interpretations because it deals with velocity and both discontinuities are frictional with the same properties at the onset of thrusting. A tentative solution was proposed here where the ramp is purely frictional and the axial surface shear-enhanced compacting. This compaction is possibly important since material is convected through the axial surface while it is always the same material in the HW, which is slipping on the ramp. Moreover, poorly compacted sediments are expected at shallow depths at the front of fold-and-thrust belts. It was shown that with these two deformation mechanisms asymmetrically applied, that indeed a hinterland thrusting ramp could become dominant if the compaction mechanism had a spatial gradient to ensure that the deformation takes place at the wedge front.

The third technical question concerns the change in decollement depth at the front of an accreting wedge. For a triangular wedge with a flat accreting front and homogeneous properties, the change from a lower to an upper decollement occurs rarely to the front, as it could be predicted with the classical CCW theory. A gradual decrease of the topographic slope at the accretion front was found beneficial in bringing the decollement activation depth change in that region.

How will the applications of LA in the earth sciences develop in the future? If one considers the number of applications of the CCW theory today and the potential of LA to go one step further, it seems that simple analytical constructions as presented in this contribution should always have a role to play. Evolution combining geometrical rules and LA are also attractive for further development either at the length scale of a single fold (Kampfer & Leroy, 2012) or at the scale of the whole accretionary wedge (Mary et al., 2013b). The difficulty, as illustrated in Section 5, is managing the mosaic of different regions from a purely geometrical point of view and from a material heterogeneity point of view in the future.

Two other avenues of future research directions should be discussed. The first concerns the restoration of the folded sequences which was proposed by Dahlstrom (1969). This is a prerequisite to any forward interpretation of thrusted sequences and is done mostly by applying geometrical rules despite the tentatives to include some mechanics (Maerten & Maerten, 2006; Moretti, Lepage, & Guiton, 2006). The restoration could be done by application of inverse methods relying on millions of forward simulations to describe completely the material property probability space. SLA offers the right level of complexity to construct the forward simulations. Such inversion approach has been considered by Cubas et al. (2013) with an application to sand-box experiments and by Caër et al. (2015) to study Mont Terri, Switzerland. The second research direction is more technical and geared toward mechanicians and it concerns first, the application of LA in 3D, a difficult task because of the construction of the pertinent virtual velocity fields and second, the account of viscous materials and thus the introduction of a time scale, another challenge from the theoretical point of view.

ACKNOWLEDGMENTS

The analytical developments presented in this paper were conducted while the first author was a resident of the Key Laboratory of Computational Geodynamics, at the University of the Chinese Academy of Sciences, Beijing, China. Prof. Yaolin Shi and Prof. Huai Zhang as well as the Chinese Academy of Sciences are thanked for their kind hospitality. The collaboration between ENS, Paris and the Chinese Academy of Sciences was initiated by the late Prof. Claude Froidevaux, and this paper is dedicated to his memory.

SUPPLEMENTARY DATA

Supplementary data related to this article can be found at http://dx.doi.org/10.1016/bs.agph.2016.09.002

REFERENCES

Baud, P., Vajdova, V., & Wong, T.-F. (2006). Shear-enhanced compaction and strain localization: inelastic deformation and constitutive modeling of four porous sandstones. *Journal of Geophysical Research: Solid Earth, 111*(B12).

Buiter, S. J. H., Babeyko, A. Y., Ellis, S., Gerya, T. V., Kaus, B. J. P., Kellner, A., ... Yamada, Y. (2006). *The numerical sandbox: Comparison of model results for a shortening and an extension experiment* (Vol. 253, pp. 29—64). Geological Society, London, Special Publications.

Caër, T., Maillot, B., Souloumiac, P., Leturmy, P., de Lamotte, D. F., & Nussbaum, C. (2015). Mechanical validation of balanced cross-sections: the case of the Mont Terri anticline at the Jura front (NW Switzerland). *Journal of Structural Geology, 75*, 32—48.

Chadwick, P. (1976). *Continuum mechanics: Concise theory and problems*. New York: Halsted Press, 174 pp.

Chen, W. (1975). *Limit analysis and soil plasticity*. Amsterdam: Elsevier.

Cooke, M. L., & Madden, E. H. (2014). Is the earth lazy? A review of work minimization in fault evolution. *Journal of Structural Geology, 66*, 334—346.

Cooke, M. L., & Murphy, S. (2004). Assessing the work budget and efficiency of fault systems using mechanical models. *Journal of Geophysical Research: Solid Earth, 109*(B10).

Crans, W., Mandl, G., & Haremboure, J. (1980). On the theory of growth faulting: a geomechanical delta model based on gravity sliding. *Journal of Petroleum Geology, 2*(3), 265—307.

Cruz, L., Malinski, J., Wilson, A., Take, W., & Hilley, G. (2010). Erosional control of the kinematics and geometry of fold-and-thrust belts imaged in a physical and numerical sandbox. *Journal of Geophysical Research: Solid Earth, 115*(B9), B09404.

Cubas, N., Barnes, C., & Maillot, B. (2013). Inverse method applied to a sand wedge: estimation of friction parameters and uncertainty analysis. *Journal of Structural Geology, 55*, 101—113.

Cubas, N., Leroy, Y., & Maillot, B. (2008). Prediction of thrusting sequences in accretionary wedges. *Journal of Geophysical Research: Solid Earth, 113*(B12).

Cubas, N., Souloumiac, P., & Singh, S. C. (2016). Relationship link between landward vergence in accretionary prisms and tsunami generation. *Geology*. http://dx.doi.org/10.1130/G38019.1.

Dahlen, F. A. (1984). Noncohesive critical Coulomb wedges: an exact solution. *Journal of Geophysical Research: Solid Earth, 89*(B12), 10125—10133.

Dahlstrom, C. (1969). Balanced cross sections. *Canadian Journal of Earth Sciences, 6*(4), 743—757.

Davis, D., Suppe, J., & Dahlen, F. A. (1983). Mechanics of fold-and-thrust belts and accretionary wedges. *Journal of Geophysical Research: Solid Earth, 88*(B2), 1153—1172.

Ellis, S., Schreurs, G., & Panien, M. (2004). Comparisons between analogue and numerical models of thrust wedge development. *Journal of Structural Geology, 26*(9), 1659—1675.

Hafner, W. (1951). Stress distributions and faulting. *The Geological Society of America Bulletin, 62*(4), 373—398.

Hardy, S., Duncan, C., Masek, J., & Brown, D. (1998). Minimum work, fault activity and the growth of critical wedges in fold and thrust belts. *Basin Research, 10*(3), 365—373.

Kampfer, G., & Leroy, Y. (2009). Imperfection and burial-depth sensitivity of the initiation and development of kink folds in laminated rocks. *Journal of the Mechanics and Physics of Solids, 57*(8), 1314—1339.

Kampfer, G., & Leroy, Y. (2012). The competition between folding and faulting in the upper crust based on the maximum strength theorem. *Proceedings of the Royal Society A*. http://dx.doi.org/10.1098/rspa.2011.0392.

Konstantinovskaya, E., & Malavieille, J. (2011). Thrust wedges with décollement levels and syntectonic erosion: a view from analog models. *Tectonophysics, 502*(3), 336—350.

Krabbenhoft, K., & Damkilde, L. (2003). A general non-linear optimization algorithm for lower bound limit analysis. *International Journal for Numerical Methods in Engineering, 56*(2), 165–184.
Krabbenhøft, K., & Lyamin, A. (2014). *Optum g2. Optum computational engineering.*
Lehner, F. K. (1986). Comments on "noncohesive critical Coulomb wedges: an exact solution" by F. A. Dahlen. *Journal of Geophysical Research: Solid Earth, 91*(B1), 793–796.
Maerten, L., & Maerten, F. (2006). Chronologic modeling of faulted and fractured reservoirs using geomechanically based restoration: technique and industry applications. *AAPG Bulletin, 90*(8), 1201–1226.
Maillot, B., & Leroy, Y. M. (2003). Optimal dip based on dissipation of back thrusts and hinges in fold-and-thrust belts. *Journal of Geophysical Research: Solid Earth, 108*(B6).
Maillot, B., & Leroy, Y. M. (2006). Kink-fold onset and development based on the maximum strength theorem. *Journal of the Mechanics and Physics of Solids, 54*(10), 2030–2059.
Malvern, L. (1969). *Introduction to continuum mechanics* (p. 320). Engle Cliffs, NJ: Prentice Hall Inc.
Mary, B. C., Maillot, B., & Leroy, Y. M. (2013a). Deterministic chaos in frictional wedges revealed by convergence analysis. *International Journal for Numerical and Analytical Methods in Geomechanics, 37*(17), 3036–3051.
Mary, B. C., Maillot, B., & Leroy, Y. M. (2013b). Predicting orogenic wedge styles as a function of analogue erosion law and material softening. *Geochemistry, Geophysics, Geosystems, 14*(10), 4523–4543.
Masek, J. G., & Duncan, C. C. (1998). Minimum-work mountain building. *Journal of Geophysical Research: Solid Earth, 103*(B1), 907–917.
Moretti, I., Lepage, F., & Guiton, M. (2006). KINE3D: a new 3D restoration method based on a mixed approach linking geometry and geomechanics. *Oil and Gas Science and Technology, 61*(2), 277–289.
Ortiz, M., Leroy, Y., & Needleman, A. (1987). A finite element method for localized failure analysis. *Computer Methods in Applied Mechanics and Engineering, 61*(2), 189–214.
Pons, A., & Leroy, Y. (2012). Stability of accretionary wedges based on the maximum strength theorem for fluid-saturated porous media. *Journal of the Mechanics and Physics of Solids, 60*(4), 643–664.
Pons, A., Leroy, Y. M., & Lallemant, S. (2013). Fluid pressure control on splay fault activation in accretionary prism based on the maximum strength theorem with application to the Nankai wedge. *Earth and Planetary Science Letters, 368*, 43–50.
Price, R. (1986). The southeastern Canadian Cordillera: thrust faulting, tectonic wedging, and delamination of the lithosphere. *Journal of Structural Geology, 8*(3), 239–254.
Rudnicki, J. W. (2004). Shear and compaction band formation on an elliptic yield cap. *Journal of Geophysical Research: Solid Earth, 109*(B3).
Rudnicki, J., & Rice, J. (1975). Conditions for the localization of deformation in pressure-sensitive dilatant materials. *Journal of the Mechanics and Physics of Solids, 23*(6), 371–394.
Salençon, J. (2002). *De l'élasto-plasticité au calcul à la rupture.* Paris: École Polytechnique, Palaiseau, and Ellipses.
Salençon, J. (2012). *Handbook of continuum mechanics: General concepts thermoelasticity*. Springer Science & Business Media.
Sans, M., Muñoz, J., & Vergés, J. (1996). Triangle zone and thrust wedge geometries related to evaporitic horizons (southern Pyrenees). *Bulletin of Canadian Petroleum Geology, 44*(2), 375–384.
Siavelis, M., Guiton, M. L. E., Massin, P., & Moës, N. (2013). Large sliding contact along branched discontinuities with X-FEM. *Computational Mechanics, 52*(1), 201–219.

Souloumiac, P., Leroy, Y. M., Maillot, B., & Krabbenhøft, K. (2009). Predicting stress distributions in fold-and-thrust belts and accretionary wedges by optimization. *Journal of Geophysical Research: Solid Earth, 114*(B9), B09404.

Stockmal, G. S., Beaumont, C., Nguyen, M., & Lee, B. (2007). Mechanics of thin-skinned fold-and-thrust belts: insights from numerical models. *Geological Society of America Special Papers, 433*, 63–98.

Strayer, L. M., & Hudleston, P. J. (1997). Numerical modeling of fold initiation at thrust ramps. *Journal of Structural Geology, 19*(3), 551–566.

Suppe, J. (1983). Geometry and kinematics of fault-bend folding. *American Journal of Science, 283*(7), 684–721.

Tarantola, A. (2005). *Inverse problem theory and methods for model parameter estimation.* SIAM.

Triantafyllidis, N., & Leroy, Y. (1997). Stability of a frictional, cohesive layer on a viscous substratum: validity of asymptotic solution and influence of material properties. *Journal of Geophysical Research: Solid Earth, 102*(B9), 20–551.

Varsek, J. (1996). Structural wedges in the Cordilleran crust, southwestern Canada. *Bulletin of Canadian Petroleum Geology, 44*(2), 349–362.

Wong, T.-F., Baud, P., & Klein, E. (2001). Localized failure modes in a compactant porous rock. *Geophysical Research Letters, 28*(13), 2521–2524.

Yuan, X., Leroy, Y., & Maillot, B. (2015). Tectonic and gravity extensional collapses in overpressured cohesive and frictional wedges. *Journal of Geophysical Research: Solid Earth, 120*(3), 1833–1854.

CHAPTER TWO

Near-Surface Geophysical Characterization of Areas Prone to Natural Hazards: A Review of the Current and Perspective on the Future

A. Malehmir[*,1], L.V. Socco[§], M. Bastani[¶], C.M. Krawczyk[||,***], A.A. Pfaffhuber[#], R.D. Miller[**], H. Maurer[§§], R. Frauenfelder[§§§], K. Suto[¶¶], S. Bazin[§§§], K. Merz[||||], T. Dahlin[##]

[*]Uppsala University, Uppsala, Sweden
[§]Politecnico di Torino, Torino, Italy
[¶]Geological Survey of Sweden, Uppsala, Sweden
[||]GFZ German Research Centre for Geosciences, Potsdam, Germany
[#]Norwegian Geotechnical Institute, Perth, Australia
[§§§]Norwegian Geotechnical Institute, Oslo, Norway
[**]Kansas Geological Survey, Lawrence, Kansas, United States
[§§]ETH Zurich, Zurich, Switzerland
[¶¶]Terra Australis Geophysica Pty ltd, Brisbane, Australia
[||||]Geo2X, Baar, Switzerland
[##]Lund University, Lund, Sweden
[***]Technical University of Berlin, Berlin, Germany
[1]Corresponding author: E-mail: Alireza.Malehmir@geo.uu.se

Contents

Advances in Geophysics, Volume 57
ISSN 0065-2687
http://dx.doi.org/10.1016/bs.agph.2016.08.001

Abstract

Natural hazards such as landslides, floods, rockfalls, earthquakes, volcanic eruptions, sinkholes, and snow avalanches represent potential risks to our infrastructures, properties, and lives. That potential will continue to escalate with current and continued human encroachment into risk areas. With the help of geophysical techniques many of those risks can be better understood and quantified, thereby minimized and at least partly mitigated through accurate, site-specific, and proper planning and engineering. On occasions these hazards simply cannot be avoided, but better characterization and therefore understanding of the subsurface geology and natural processes responsible for the threats is possible through integration of various cost-effective geophysical methods with relevant geotechnical, geomechanical, and hydrogeological methods. With the enhanced characterization possible when geophysics is incorporated into natural hazard analysis, potential risks can be better quantified and remediation plans tuned to minimize the threat most natural hazards present to civilizations. In this article we will first review common geophysical methods that can be and have been utilized in studying natural hazard prone areas, then we provide selected case studies and approaches using predominantly our own examples, and finally a look into the future detailing how these methods and technologies can be better implemented and thereby more time- and cost-effective and provide improved results.

1. INTRODUCTION

Although there are no reliable statistics on how many geophysicists work on problems related to natural hazards or the near surface in general, it is likely approaching the number involved in natural resource exploration and exploitation. More than 3000 members of EAGE (European

Association of Geoscientists and Engineers) and SEG (Society of Exploration Geophysicists) have registered their primary field of expertise as near-surface geosciences and related matters (Laurie Whitesell, personal communication, April 2016). In spite of the large number of investigations routinely carried out in geohazard prone areas, natural hazards in general and geohazards in particular, such as landslides, earthquakes, and volcanic eruptions still take human lives in shocking numbers and routinely result in significant property loss, increased risk to key infrastructures, and threats to the environment.

According to a report by the United Nation, natural disasters have claimed more than 600,000 human lives and left behind trillions of dollars in damages in the last two decades (Wahlstrom & Guha-Sapir, 2015). Most of these disasters are related to extreme weather (e.g., storms, floods, landslides, and extreme heat periods). In terms of economic impact, based on around 30,000 events occurring during the period from 1900 to 2015, Daniell, Wenzel, and Schaefer (2016) reported 6.5−14 trillion dollars was lost globally; over 40% due to flood/rainfall, 26% due to earthquake, 19% due to storm effects, 12% due to drought, 2% due to wildfire, and under 1% due to volcano. While some geohazards are synonymous with and actually require settings unique to particular regions (e.g., volcanoes near subduction zones or active faults near plate boundaries), others occur in many places indiscriminate of geologic settings and normally *know no border* (Geoscientists *Without* Borders, 2016). Slope stability represents one of the more common geohazards with landslides, rockfalls, and debris flows all resulting from unstable rock or soil.

Mobile ground cover is common in most countries with steep surface topography (Figs. 1−3). According to our own Internet search, landslides alone during the second half of 2015 accounted for more than 300 lives lost in only three landslides, one in Myanmar (90 human causalities, November 2015), another in Guatemala (131 human causalities, October 2015), and a third one in China (91 human causalities, December 2015). At least one of these slides has been suggested to be the direct result of human activity (the one in China due to extreme waste construction material pilling with no or little geotechnical reinforcement) with extreme weather conditions also potentially contributing to the instability (Wikipedia, 2015).

Landslides are one of the largest global geohazards in terms of surface area potentially threatened by slope stability (Fig. 1). For example, in the United States, landslides have been reported in every state and roughly responsible for 1−2 billion dollars in damage and an average of more than 25 fatalities annually (Shan, Bastani, Malehmir, Persson, & Lundberg, 2016; USGS, 2013). The Oso landslide of 2014 (Keaton et al., 2014; Fig. 1A) claimed

Figure 1 Examples of heavy rainfall–related landslides (gravitational-related) from (A) the Oso, Washington, United States, 2014 *(photo by Mark Reid from USGS.)* and (B) the Nova Fiburgo, Rio de Janeiro, Brazil, 2011 *(photo by Nelson F. Fernnandes.)* illustrating the extent of the damage to the people, environment, and infrastructures.

43 lives, blocked a river, and caused transportation delays. From 2004 to 2010 more than 2620 landslides claimed in excess of 32,000 lives worldwide (Petley, 2012). Examples such as these have led many countries to establish landslide hazard programs to tackle the threat as a long-term issue and provide a wide range of mitigation strategies (see Shan et al., 2016 and references therein). Guidelines are generally available that describe appropriate remote sensing technologies applicable for monitoring different types of landslides (e.g., SafeLand, 2010, a European initiative), while an evaluation report on innovative monitoring and remote sensing methods with a discussion of future technologies can be found in SafeLand (2011). SafeLand (2012) reviews feasible techniques for landslide detection, fast characterization, rapid mapping, and long-term monitoring.

Figure 2 Examples of typical liquefaction-type quick clay landslides from (A) Kattmarka, mid-Norway, 2009 *(courtesy of Norwegian Geotechnical Institute, NGI.)* and (B) Småröd, southwest Sweden, 2006 *(photo by Per Petersson, previously at the Swedish Transport Administration, Trafikverket.)*. These types of slides often do not require steep topography and can occur even in gentle topography. Understanding the subsurface geology and hydrogeological regime are important to reduce the risks associated with these slides. Several case studies presented in this paper are related to these types of slides.

While weather-related hazards likely have the most significant long-term (until next glaciation time; >50,000 years or so) impact, earthquakes and volcanoes are generally credited for taking more lives and damaging more infrastructures, particularly in and around populated metropolitan areas. Volcanic eruptions, and associated hazards such as landslides and rockfalls, are often predictable due to an associated increased seismicity. A volcano can, however, unexpectedly erupt with no precursors to indicate an eminent eruption. For example, the September 2014 Mount Ontake eruption in

Figure 3 (A) Debris flow associated with the flooding and massive numbers of landslides in Teresopolis Posse, Rio de Janeiro, Brazil, 2011 *(photo by Nelson F. Fernnandes.)* and (B) rockfall associated with the same event near a city hospital *(photo by Alireza Malehmir.)*. The hospital had to be closed during the devastating 2011 landslide because of the high risk of further rockfalls.

Japan occurred without warning and claimed more than 50 human lives (Independent, 2014).

Fresh volcanic ash (extremely unconsolidated) combined with extreme rain events can result in supersaturation and mass flow or mud slides (debris flow), which have been responsible for historical causalities in countries such as Guatemala and Indonesia. The 2010 eruptions of Eyjafjallajökull volcano in Iceland gave volcanic-related hazards a different perspective. Europe exercised its first volcanic ash crisis (Bye, 2011). Ash covered large areas of northern Europe when the volcano erupted. About 20 countries closed their airspace to commercial jet traffic and it affected about 10 million travelers. In April, when the situation was quite severe, the lost revenue estimated for only three days was in the order of 400 million dollars per day (Bye, 2011).

Earthquakes are also unpredictable hazards that have been responsible for extreme loss of life and untold damage to infrastructures and properties. Some estimates suggest that more than 2 million people have been killed as direct and indirect result of earthquakes since the beginning of the last century (USGS, 2014). Blind faults have been responsible for unexpected ground motion in the past and will continue to do so. Therefore, national programs are in place to promote subsurface delineation and characterization where faults are suspected or inferred. The Bam 2003 earthquake in Iran, resulted in more than 26,000 fatalities and was due to the rupture of a blind strike-slip fault (inferred later by satellite imagery data) that extended under the historical city and was previously not known (Talebian et al., 2004).

Seismogenic models can be quite complex, and even in areas with recognized high seismicity, where investigations have been ongoing for many years, scientific debate trying to definitively identify active faults and source mechanisms is ongoing. In spite of the fact that earthquakes are known to occur in regions of active tectonics, they cannot be predicted with sufficient timing accuracy to prevent loss of life, a problem complicated by the tendency of active fault zones to be at highly populated places. For example, the April 2015 Nepal earthquake claimed more than 8000 lives and severely damaged property and infrastructures (Kobayashi, Morishita, & Yarai, 2015).

Other types of recognized geohazards include rockfall (Fig. 3B), tsunamis, sinkholes and associated subsidence, and avalanches in the glacial/alpine regions. Risks associated with avalanches can occasionally be higher than normally estimated in certain settings and under specific triggers. The earthquake-triggered 1970 Ancash avalanche in Peru took more than 20,000 lives across an approximately 18 km advanced, with an inferred speed of 250–300 km/h and displacement of an estimated 80 million m^3 of ice, snow, water, and debris (Cluff, 1970).

Natural hazards and their investigation using geophysical methods are not limited to land. Marine landslides have been reported offshore, some triggered by movement on blind faults and have produced tsunamis (e.g., Hornbach et al., 2010; Hornbach, Mann, Frohlich, Ellins, & Brown, 2011). Chirp high-resolution (3.5 kHz) submarine seismic imaging and sediment coring techniques have been used (Hornbach et al., 2011) to map active faults and recent movement on submarine slumps in and around Kingston Harbor in Jamaica. The fault is inferred to be the western terminus of the Enriquillo Plantain Garden Fault (EPGF), the same fault that ruptured and resulted in the 2010 Haiti earthquake, destroying Port-au-Prince and resulting in about 250,000 causalities (Hornbach et al., 2011). While offshore geohazards can be problematic and result in damage to infrastructures, human casualties, and environmental issues, they have generally been regarded as minor problems and therefore less attention has been given to them to date.

Geophysical methods, especially when combined with other engineering and sampling methods are capable of characterizing and categorizing regions prone to geohazards in very practical and useful detail. In spite of the number of successful examples where geophysical methods have been employed for geohazard characterization (e.g., Bogoslovsky et al., 1977), unfortunately geophysical methods are not routinely integrated into investigation schemes for various reasons. Fortunately, there is a variety and an encouraging number of initiatives promoting and encouraging the use of geophysical methods to identify threats and mitigate risks associated with natural hazards (e.g., Jacobs, 2008, 2015). However, these initiatives are not proportional to the size and diversity of the problem and the need for expanded application of geophysics.

In this review article, we aim at providing some basic information about the more commonly used geophysical methods; all that can be and most that have been applied to study areas with an elevated risk of some form of failure related to a geohazard. It is also our intent to provide examples of various types of geohazards and areas (Fig. 4) that possess some form of geohazard

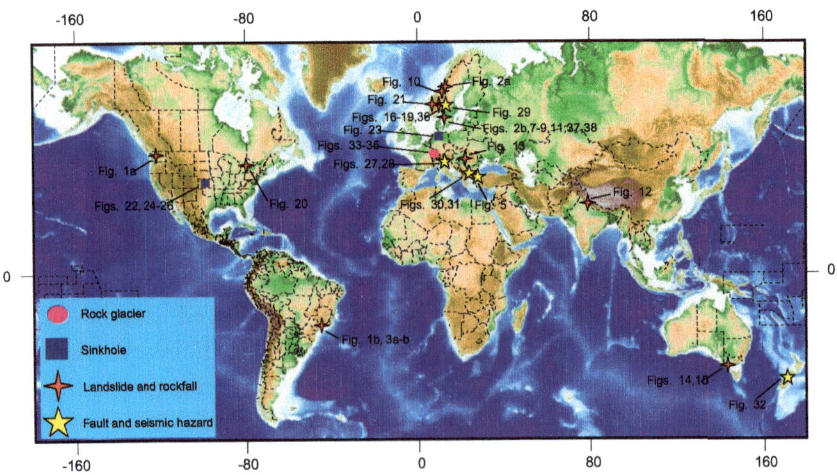

Figure 4 Map showing location of examples or photos presented in this paper.

risk; biased, of course, towards our own experiences and works. We will also look into the future and discuss potential improvements, particularly in regard to ways to acquire higher quality data and in a more time- and cost-effective manner.

2. METHODS AND THEIR BASIS

2.1 Downhole, Surface, Air- and Shipborne Surveys

Geophysical surveys can be configured in various ways, and parameters can be set up to address a wide range of objectives at a desired resolution and level of confidence. Most geophysical methods use surface measurements with sensors deployed on a temporary basis (some are done on a more semipermanent basis for time-lapse studies or monitoring; e.g., Hare, Ferguson, Aiken, & Brady, 1999) on the ground surface and in a minimally invasive fashion. There is also a specialized subset of geophysical methods designed to be employed using air- or ship-based platforms. As well, many methods lend themselves to acquisition in a borehole where these downhole measurements generally include direct physical property measurements, not routinely done for near-surface applications, where the main goal is improving signal-to-noise ratio, vertical resolution, or for validation and calibration of surface or airborne geophysical results (Labo, 1986; Sharma, 1997 and references therein; Jones, 1999; Keary et al., 2002 and references therein; Burger, Sheehan, & Jones, 2006 and references therein; Reynolds, 2011 and references therein; Everett, 2013 and references therein).

In cases where the target is in unconsolidated sediments, downhole physical property measurements can be extremely challenging since casing (even PVC-type) would be necessary, and this would limit the amount and type of information to be gathered; natural gamma is possible and often quite useful. Sampling and laboratory measurements complement the downhole logging measurements (Salas-Romero et al., 2016). Surface-downhole measurements can be quite useful but have only seen limited use due to the high cost of drilling (e.g., Cosentino, Martorana, Perniciaro, & Terranova, 2003; Jeppson, Bradbury, & Evans, 2010). Airborne, especially new unmanned aircraft systems are quite popular because they can cover a large and difficult area in a short period of time (e.g., D'Alessandro &Martorana, 2015; Whitehead & Hugenholtz, 2014) and likely to be employed extensively in the near future for hazard mapping and assessments.

2.2 Electrical Methods

The electrical resistivity of the subsurface is principally related to porosity, saturation of the pore space with conductive fluids, clay content, and metallic minerals and their spatial distribution and connection to each other. In most cases, unconsolidated sediments are less resistive than competent crystalline rocks, whereas other rocks like shales and graphite are highly conductive. Resistivity measurements can thus be a valuable method to map weak zones, sliding planes, saturated or air-filled voids, and clay occurrence as well as presence of highly sensitive quick clays. Measurements can be done using marine, ground, crosshole, and airborne platforms.

In terms of penetration depth, it can range from a few meters to hundreds of meters depending on the array type and length as well as current or frequency used. In conductive medium, resolution and penetration depth is generally less than that of resistive medium and imaging under highly conductive materials (e.g., graphite or thick column of fully saturated clay materials) can be quite challenging if not impossible (Reynolds, 2011 and references therein). Here, we will highlight a selection of the many resistivity methods used for geohazard applications.

2.2.1 Galvanic Geoelectrical Methods

Electrical resistivity tomography (ERT) is a ground-based method that uses galvanic coupled electrodes to inject current and measure potential difference at locations between or outside the current electrodes, depending on the array type used (e.g., Zonge, Wynn, & Urquhart, 2005; Ward, 1990). ERT uses a DC current with potentials measured while the current is on. During the last two decades, two-dimensional (2D) ERT surveys have grown into a well-established, standard tool for engineering and environmental applications. To a large degree this is the result of the development of user-friendly automated multielectrode data acquisition systems and easy to use PC-based inversion software (Dahlin, 2001; Loke, Chambers, Rucker, Kuras, & Wilkinson, 2013). ERT is typically carried out using layouts that consist of tens of electrodes arranged in a line, logging several hundreds or thousands of data points.

A 3D sampling distribution of resistivity measurements could be obtained either by combining a set of 2D surveys or by undertaking a real 3D data acquisition that uses 2D surface grids of electrodes. Modern ERT instruments measure potentials at several locations in parallel using multi-electrode and multichannel systems, which allow rapid data acquisition (e.g., Dahlin & Zhou, 2006). Resistivity models are typically derived for 2D or 3D cases by inverse modeling of measured data (least-square or Occam's type methods). There are several commercial software packages as well as research software programs available (e.g., Auken, Pellerin, Christensen, & Sørensen, 2006; Loke et al., 2013) for modeling and data handling. As a rule of thumb for standard nested or dipole—dipole type arrays, the near-surface resolution is approximately half the electrode spacing and the maximum depth of penetration is 10—20% of the total layout length. The resistivity distribution of the subsurface (earth properties), to a limited extent, determines the penetration depth. This is due to current channeling in low-resistivity materials, which implies, for example, that the resolution is limited in high-resistivity materials beneath a good conductor.

In addition to only electrical resistivity measurements, most instruments can continue taking measurements during the current off-time, which provides information on the chargeability of the subsurface materials. This approach is commonly referred to as time-domain-induced polarization or IP method. This combined measurement approach is also referred to as DCIP tomography (direct current resistivity and time-domain IP). Some ERT and IP systems apply an AC current across a range of frequencies to determine chargeability as a function of frequency. Chargeability can be an indicator of clay or shale in the subsurface or disseminated metallic mineralization as well as groundwater contamination (e.g., Zong et al., 2005).

Time-domain IP has traditionally been designated as integral chargeability over a relatively long decay interval. However, recent developments have opened up possibilities to extract spectral IP parameters from time-domain IP data. This approach provides additional useful information related to the internal structure of subsurface geological materials (Fiandaca et al., 2013; Hönig & Tezkan, 2007). It is, however, important to acknowledge that from a data quality point of view, IP data are much more challenging to acquire than resistivity data, because of the propensity for coupling problems in the measurement spread and smaller signal levels increasing the sensitivity to noise (e.g., Dahlin & Leroux, 2012; Gazoty, Fiandaca, Pedersen, Auken, & Christiansen, 2013).

ERT can be applied as a mapping or monitoring tool, with the latter believed to pose a large opportunity for natural hazard risk assessment. By analyzing time-series data from temporary or permanently installed electrode arrays data relating to variations in water saturation, temperature, gas content, or frozen ground can be recovered. These kinds of data can yield valuable early warning information in connection with landslide risk, anomalous water leakage, gas outflow, and permafrost degradation, to name a few (e.g., deBari, Lapenna, Perrone, Puglisi, & Sado, 2011; Buchler et al., 2004; Chambers et al., 2009; Miller, Routh, Brosten, McNamara, 2008; Miller, Villella, Xia, 2008). Expansion of the concept to DCIP monitoring might enhance the information for some applications.

2.2.2 Inductive Electromagnetic Methods

While the ERT method establishes a galvanic contact to the ground through direct and normally invasive electrodes, electromagnetic (EM) methods make use of the eddy currents induced in the ground both from sources (active or passive) on/near the ground or towed by aircrafts or helicopter. The primary field (the source of the EM signal) induces an electrical current in the earth media, which in turn, causes a secondary EM field. The secondary EM field is generally measured through the rate of change in the magnetic flux within a circuit loop that does not have direct contact with the earth. It can be combined with electric field measurements made through contact with the ground (electrodes) for higher lateral resolution.

The primary EM field can be natural (originated from solar winds or lighting) or artificial with EM transmitters (local or at distances like ordinary radio transmitters used in the navigation industry). There are many techniques and variations to the EM method (for example, time versus frequency domains; e.g., Zhdanov, 2009 and references therein; Dentith & Mudge, 2014 and references therein) but common in all these variations is measurement of distortions of the primary EM field, which is related to the electrical resistivity of the subsurface materials.

One of the less frequently used ground-based EM methods discussed in this paper is radio magnetotellurics (RMT). RMT is a frequency domain plane-wave EM method. The sources of the EM signals are distant radio transmitters operating in the 15–250 kHz frequency range. These transmitters are mainly used for navigation purposes (Pedersen, Bastani, & Dynesius, 2006). At each RMT station three components of the magnetic field and two horizontal components of the electric field are registered at the same time (Bastani, Persson, Mehta, & Malehmir, 2015). Data are frequency-dependent; impedance tensor elements, and tipper vector, which provide valuable information about the electrical resistivity structures below the measuring points. The penetration depth of the method can vary between 5 and 500 m and depends on the ratio between electrical resistivity of the ground and the signal frequency (known as the skin depth). Field data are usually presented as variation of apparent resistivity and impedance phases versus transmitter frequencies at each station or along a profile. Depending on the acquisition geometry, RMT data can be inverted in 1D, 2D, or 3D (Bastani et al., 2012; Shan et al., 2016). Kalscheuer et al. (2013), for example, present results from single and joint 2D inversions of ERT, and controlled-source audio-magnetotelluric data along a single profile targeting a quick-clay landslide site in Norway.

Airborne electromagnetics (AEM) is an efficient and rapid way to acquire data for imaging the subsurface distribution of resistivity from rotary or fixed-wing aircrafts. Adopted originally from the mineral exploration industry in the 1950s (Fountain, 1998), over the last 15 years major improvements in hardware as well as software have rendered AEM an accurate and precise near-surface geophysical method (e.g., Sørensen & Auken, 2004). Helicopter time-domain EM (known as TEM) systems are now routinely used for near-surface applications. A large loop enclosing several hundred square meters with less than 10 wire turns sends an impulse with several 100,000 Amp.m^2 moment into the ground, inducing eddy currents, which in turn create a secondary magnetic field. This secondary field is recorded during the off-time (after the transmitter switched off the current) with the measured time decay inverted to produce a model of the subsurface resistivity. Typically, these models are calculated in 1D (Viezzoli, Christiansen, Auken, & Sørensen, 2008) and on rare occasions in 3D (Cox, Wilson, & Zhdanov, 2012). Recent work has shown that the resolution and accuracy of modern AEM are close to ground (ERT) measurements with penetration depths typically reaching several hundred meters, depending on system parameters and subsurface conductivity (Anschütz, Bazin, Kåsin, Pfaffhuber, & Smaavik, 2016). Use of AEM methods is likely to see increased attention as a tool for better planning and inventorying of geohazards, given these systems' potential for quick mapping and subsurface characterizations particularly in remote or difficult terrain conditions.

2.3 Ground-Penetrating Radar

Ground-penetrating radar (GPR) is widely used across a broad range of applications in engineering and environmental geophysics. Some of these applications include concrete assessment, hydrogeophysics, utility/excavation location, unexploded ordnance detection, and a diverse set of geological applications (e.g., Holliger & Bergmann, 2002; Linde, Binley, Tryggvason, Pedersen, & Revil, 2006; Rubin et al., 2006 and references therein; Jol, 2008 and references therein; Looms, Binley, Jensen, Nielsen, & Hansen, 2008; Miller, Ivanov, Xia, Peterie, & Sloan, 2010 and references therein; van der Kruk, Revil, & Slob, 2010 and references therein; Binley et al., 2015; Cassiani et al., 2012 and references therein). GPR is also a suitable technique for natural hazard assessments. For example, it has proved to be useful for characterizing rockslide areas (e.g., Heincke, Green, van der Kruk, & Horstmeyer, 2005), quick-clay sites (e.g., Sauvin et al., 2013), and unstable permafrost (e.g., Maurer & Hauck, 2007). GPR performs well in electrically resistive environments but less effective in saturated clays. Most surveys for geological applications employ frequencies in the range of 10–200 MHz, resulting in a maximum penetration depth of a few tens of meters and a spatial resolution in the meter to submeter range. GPR measurements can be done in a variety of configurations, for example, on the surface, in boreholes, or from helicopters.

2.4 Seismic Methods Including Seismicity

Seismic waves used for near-surface applications are usually generated by artificial, controlled sources and ground vibration and recorded with lines or grids of sensors (geophones or accelerometers). The recorded signal contains body waves, surface waves, and noise (source and cultural) and can be processed and then interpreted to provide information about seismic properties (e.g., velocity and attenuation) related to mechanical and petrophysical parameters (e.g., elastic properties, porosity, and fluid content, etc.). Hence, near-surface seismic methods are primarily used in engineering studies and as a tool in the mitigation of risks associated with potential geohazards. Seismic surveys can be performed along the ground surface, in boreholes, and underground spaces (e.g., tunnels) using various configurations (Yilmaz, 1987, 2015 and references therein; Sheriff & Geldart, 1995 and references therein).

Traditionally, seismic methods based on body wave energy have been divided into reflection and refraction (first arrival) methods. The reflection method can provide detailed subsurface geometries and variations in sediment properties when there is sufficient seismic impedance (product of seismic velocity times bulk density) contrast and favorable recording conditions and application (e.g., coupling, signal-to-noise ratio, attenuation, target size, and geometry). Refraction methods can provide interval velocities and layering information as long as there is an increasing velocity gradient with depth (Pelton, 2005). Refraction as well tomographic inversion-based methods use first or secondary arrivals and have routinely been used to study many kinds of geohazards, such as landslides (e.g., Adamczyk, Malinowski, & Malehmir, 2013, 2014; Bogoslovsky and Ogilvy, 1997; Carroll, Scott, & Lee, 1972; Wang, Malehmir, & Bastani, 2016), fault detection and delineation (e.g., Calvert, Fisher, & Ships working group, 2001; Malehmir et al., 2016; Morey & Schuster, 1999), sinkholes (e.g., Hanafy, 2010), and delineation of magma chambers in volcanic regions (e.g., Nagaoka, Nishida, Aoki, Takeo, & Ohminato, 2012; White & Clowes, 2007).

Given that the method suffers where there is no positive velocity gradient or in the case of hidden layers (low velocities below high velocity materials or surrounded by them) there are some natural hazards or regions that are not conducive to the use of refraction methods such as glaciated regions or cavities. This can be overcome in some places with reflection seismic or GPR (see later) methods (e.g., Burschil & Krawczyk, 2016; Hanafy, 2010). Penetration depth and resolution of seismic methods are generally some combination of seismic wavelength and maximum offset used for data acquisition. Typical refraction penetration depth is about 15–20% of the maximum recorded sensor offset in the data. Survey geometry such as source and receiver spacing, and offset source and receiver types, as well as medium's velocity distribution control spatial resolution in refraction seismic surveys. Generally speaking, resolution decreases with depth.

The reflection method is not routinely applied for geohazard studies with the exception of fault and fracture imaging (e.g., Malehmir, Durrheim, et al., 2012; Malehmir, Juhlin, et al., 2012; Williams et al., 2005 and references therein; Juhlin, Dehghannejad, Lund, Malehmir, & Pratt, 2010; Place, Malehmir, Högdahl, Juhlin, & Persson Nilsson, 2015), ground subsidence/stability (e.g., Sloan, Peterie, Ivanov, Miller, & McKenna, 2010), or recently even in volcanic areas (e.g., Krawczyk, Polom, Wiyono, & Pramono, 2015) where continuous reflectors are usually not expected to be present. A significant difference between seismic reflection and seismic refraction surveying is the acquisition setup, source capabilities, channel count (i.e., number of sensors), and processing requirements. Seismic reflection is more time-consuming and expensive as compared to the seismic refraction method.

The refraction method requires long offset sensors to obtain increased depth penetration while the reflection method requires denser receiver and shot spacing (mainly to increase signal-to-noise ratio) with higher fidelity sources and offset ranges that are ideally tuned for various shape and size targets. There are examples where a seismic survey is designed so that both refraction and reflection imaging are possible with the same data set due to ready accessibility of a large number of receivers to optimize the use of broadband energy sources (Krawczyk, Polom, & Beilecke, 2013; Krawczyk, Polom, Malehmir, & Bastani, 2013; Krawczyk, Polom, Trabs, & Dahm, 2012; Malehmir, Bastani et al., 2013; Malehmir, Saleem, Bastani, 2013; Malehmir, Wang, et al., 2015; Malehmir, Zhang, et al., 2015; Polom et al., 2013; Polom, Druivenga, Grossmann, Grüneberg, & Rode, 2011; Polom et al., 2010; Pugin et al., 2013; Wang et al., 2016). In terms of spatial resolution, reflection seismic methods are dominantly sensitive to the seismic wavelength and when target is smaller than a quarter of the seismic wavelength, the response would be in the form of diffraction (observable) and/or scattering (hidden in the seismic background), respectively (e.g., Cheraghi, Malehmir, & Bellefleur, 2012; Cheraghi, Malehmir, Bellefleur, Bongajum, Bastani, 2013; Malehmir & Bellefleur, 2009; Malehmir, Bellefleur, & Müller, 2010).

Most active-source refraction and reflection surveys use a vertical type source and receivers (of various natural frequency), however, this is also changing and more and more multicomponent surveys using horizontal and vertical type sensors and sources are used for geohazard studies (Krawczyk, Henk, et al., 2015; Krawczyk, Polom, & Dahm, 2012; Malehmir, Wang, et al., 2015; Malehmir, Zhang, et al., 2015; Pugin, Larson, & Sargent, 2004, 2009). For near-surface studies, horizontal component surveys (such as horizontal receivers—horizontal sources, known as SH—SH survey) are becoming more commonly employed given the simplicity of the wave type and greater resolution. Our experience shows that recording high frequency shear-wave data, as for P-wave data, is often very challenging for near-surface studies, but the fact that shear-wave velocities are much lower than P-wave velocities for the same material (occasionally reaching 10 times lower) allows high-resolution imaging of the subsurface geology at lower frequencies using SH—SH type surveys.

It is the goal of any near-surface geophysical survey to record the highest resolution data possible without compromising high signal-to-noise ratios that allow a highly confident interpretation of thin and small features. For many near-surface settings, P-wave seismic reflection surveys produce high signal-to-noise ratios but can be challenged to equal the resolution potential of an equivalent S-wave reflection survey. The historical popularity of P-wave reflection surveys has been driven by the ease of producing and recording high confidence reflections from an optimum reflection window that is void of coherent source noise (Hunter, Pullan, Burns, Gagne, & Good, 1984). With the advent of small, reliable shear wave sources and towed spreads, it has become a surge in acquiring high quality shear wave reflection data.

One of the challenges that has kept shear wave reflection profiling principally in the hands of the experts is that in many near-surface settings with thick sequences of unlithified sediments, the shear-wave and surface-wave velocities are close enough that coherent energy from both arrivals are observed on shot gathers at nearly the same time and with similar spectral properties. Separating dispersive surface-wave arrivals from hyperbolic shear-wave reflection arrivals returning from the shallow subsurface can be challenging (Kelly, 1983). It is for these reasons that most near-surface seismic reflection surveys still employ compressional wave energy, but the allure and potential of S-wave reflection surveying has not gone unnoticed by most practitioners.

In the last two decades, besides classical body-wave surveying, surface-wave analysis has gained great popularity, particularly in engineering and for geohazard studies (see Miller, Xia, Park, & Ivanov, 1999; Socco, Foti, & Boiero, 2010; Socco, Jongmans, et al., 2010; Socco, Boiero, et al., 2010; Foti, Parolai, Albarello, & Picozzi, 2011 and references therein). Surface waves (mainly Rayleigh waves but also Love and Scholte waves) in layered media exhibit a dispersive behavior; their phase velocity depends on the wavelength (or frequency) and is related to the seismic velocity distribution at the site (Park, Miller, & Xia, 1999). Since surface waves are particularly sensitive to S-wave velocity, and relatively insensitive to P-wave velocity, the unknown parameters in the inversion are usually S-wave velocity and layer thicknesses while P-wave velocity (or Poisson's ratio) are assumed *a priori* (Xia, Miller, Park, Hunter, & Harris, 2000).

The inversion of surface-wave estimated phase velocities is normally performed assuming a layered system but several approaches have been proposed to focus the investigation and achieve significant lateral resolution in laterally varying environments (Bergamo, Boiero, & Socco, 2012; Tatsunori, Tsuji, & Matsuoka, 2013). Since the investigation depth depends on the maximum wavelength of the recorded surface waves, active data are often complemented with passive noise acquisitions that allow a broadening of the frequency bandwidth toward the low frequencies (see for instance Foti, Lai, Rix, and Strobbia (2014) for a comprehensive description of the method).

The popularity enjoyed by surface-wave analysis lies in the ease of estimating S-wave velocity models as function of depth. S-wave velocity is a key parameter in several geohazard studies; for instance, seismic site response mainly depends on S-wave velocity distribution of near-surface layers and S-wave impedance contrasts. Bour, Fouissac, Dominique, and Martin (1998), Parolai and Galiana-Merino (2006), and Hunter and Motezedian (2006) illustrate examples of providing standards for seismic noise analysis in urban areas resulting in shear-wave velocity maps for subsurface characterization. Other examples include landslides, settlement, and subsidence problems requiring knowledge of shear modulus of materials for engineering purposes.

Nevertheless, very often surface-wave dispersion is extracted from seismic records gathered for seismic reflection or seismic refraction surveys; it is hence possible to perform joint inversion of dispersion curves and body wave traveltimes to obtain internally consistent and better constrained velocity models (Boiero & Socco, 2014; Re, Strobbia, De Stefano, & Virgilio, 2010). Borehole information can also be included as *a priori* information and local constraints (Socco, Boiero, Foti, & Wisén, 2009). These approaches, particularly when complemented by structural information coming from high-resolution reflection seismic data provide a comprehensive subsurface model of the mechanical properties of the solid matrix (S-wave velocity model) and of fluids distribution (P-wave model) (e.g., Kostantakis et al., 2013).

Passive seismic or microtremor surveys (Okada, 2003) are commonly used to investigate S-wave velocity structures in a potential hazard at a medium to large scale. This approach is used to identify areas where earthquake motion may be amplified by geological structure. Horizontal to vertical spectrum ratio (H/V) of microtremors also provides a basis for geohazard study (Claprood & Asten, 2009; Picozzi et al., 2010). While still at the research and field demonstration stage, passive seismic data (and their combinations with active source data) are becoming popular for geohazard studies such as in areas near volcanoes and active faults (e.g., Bazin et al., 2010) or where rockfall is possible (e.g., Bottelin et al., 2013). Passively recorded seismic data can be used to locate seismically active zones as well as to derive P- and S-wave velocities to study zones of weaknesses in unconsolidated sediments potentially susceptible to landslide or the formation of sinkholes (e.g., Kühn, Ohrnberger, Vollmer, & Dahm, 2011).

2.5 Potential Field (Gravity and Magnetic) Methods

Potential field methods are among the oldest geophysical methods and originally introduced for resource exploration. Their time and cost effectiveness (particularly if acquired from an airborne platform) allows them to be feasible for geohazard applications. Here we refer to potential fields mainly as gravity and magnetic methods, which require a density or susceptibility contrast in the subsurface to be successfully applied to an imaging or characterization problem. In terms of resolution, these methods have high sensitivity to near-surface anomalies and loose their resolution significantly with depth, hence resolution is also reduced when used on an airborne platform.

Potential field measurements can be performed on the ground surface, in boreholes, or as air- and shipborne (e.g., Hinze, von Frese, & Saad, 2013 and references therein). Magnetic data typically have higher lateral resolution due to larger susceptibility contrasts than gravity data. In both cases, usually there are a number of equally valid models that can describe the data (issues with nonuniqueness and ambiguity in their modeling).

Gravity methods have successfully been used to study sinkholes (e.g., Eppelbaum, Ezersky, Al-Zoubi, Goldshmidt, & Legchenko, 2008; Paine, Buckley, Collins, Wilson, & Kress, 2009; Rybakov, Goldshmidt, Fleischer, & Rotstein, 2001), cavities (e.g., Butler 1984), and magma activities. Continuous gravity and magnetic measurements combined with geotemperature data were used to monitor the Akita-Komagatake volcano in Japan from 1970 until 2014 (Kitsunezaki & Muraoka, 2015). A good correlation was observed between geothermal activities and gravity and magnetic field decrease and vice versa.

Most faults show good magnetic signature either as minima or maxima (e.g., Dehghannejad, Bauer, Malehmir, Juhlin, & Weihed, 2012; Malehmir et al., 2011, 2016) depending on the fluid activities within the fault (e.g., hematization of magnetite or magnetite forming during the fluid circulation). Faults can also possess similar signatures on gravity data, but much smoother in terms of character. Airborne magnetic measurements have also been used to study hydrothermally altered rocks that can weaken volcanoes leading to their collapse or destructive debris flow (Finn, Sisson, & Deszcz-Pan, 2001).

For landslide studies, potential field data are effective in mapping bedrock undulations and possibly slip surfaces. Salas-Romero et al. (2016), for example, found that a slip surface for quick-clay slide can be coarse-grained materials, and in most cases these materials showed strong magnetic susceptibility indicating the presence of magnetite in the sand. While stratiform in this case, it may be possible to use magnetic measurements to study their lateral variations.

Time-lapse gravity measurements are performed to monitor mass movements in the subsurface, with applications ranging from groundwater movement (e.g., Davis, Li, & Batzle, 2008; Naujoks, Weise, Kroner, & Jahr, 2008), via volcanic activity (e.g., Jentzsch et al., 2001) to geological CO_2 storage (Arts, Chadwick, Eiken, Thibeau, & Nooner, 2008). In the Dead Sea region (Rybakov et al., 2001) and across relict salt mines in the UK (Pringle et al., 2012), gravimetric monitoring has proven successful for imaging regions of subsidence. With subsequent data modeling, it is possible to produce progress scenarios for vertical migration of voids or collapse events.

2.6 Interferometric Synthetic Aperture Radar

Interferometric synthetic aperture radar (InSAR) is a radar technique used principally in geodesy and remote sensing applications. It uses two or more synthetic aperture radar (SAR) images, acquired at different times to generate maps of surface deformation. The method utilizes differences in the phase of the waves returning to the satellite or aircraft (e.g., Bamler & Hartl, 1998; Massonnet & Feigl, 1998; Rosen et al., 2000). The technology can be used for studying hidden faults, surface deformation, volcanic activities, landslides, and for structural engineering, in particular for monitoring rockfall, subsidence, and structural stability. It can also be used to estimate total mass movement, which allows a quick evaluation of a slide, for example. The major limiting factors for the use of the method arises in mountainous and/or densely vegetated terrains where decorrelation and complex SAR image geometry can lead to incomplete spatial coverage. Unlike other remote sensing devices, radar sensors are not particularly influenced by weather and illumination.

Unlike satellite-based InSAR, ground-based InSAR employs a fixed placement of the measurement equipment at a site. This reduces the sources of error in the measurements but increases the acquisition time. The technology has been in use since the early 2000s, and can be used as an individual monitoring tool for, e.g., exposed rock faces or as a useful supplement to different existing monitoring systems (e.g., Monserrat, Crosetto, & Luzi, 2014 and references therein).

2.7 Lidar

Lidar is a surveying technology that measures distance by illuminating a target with a laser light (Goyer & Watson, 1963). Lidar is a popular technology for making high-resolution topographic maps with a wide range of (e.g., Gutiérrez et al., 2001 and references therein) actual and potential airborne and terrestrial applications (Vosselman & Maas, 2012). Different applications and specific use of lidar data require various scanner specifications, which include the size of the area to be captured and the range of measurements desired that in turn dictate the cost of equipment.

Airborne lidar utilizes laser scanners attached to a plane with acquisition occurring during flight, and thereby creating a 3D point cloud model of the landscape. This is currently the most detailed and accurate method of creating digital elevation models, replacing photogrammetry. Lidar products and services are readily available in most developed countries (e.g., Mikko, Smith, Lund, Ask, & Munier, 2015). One major advantage of lidar in comparison with photogrammetry is the ability to filter out vegetation from the point cloud model in route to create a digital surface model. In this fashion, areas covered by vegetation can be visualized and the ground surface mapped, including rivers, paths, cultural heritage sites, as well as others targets of interest. Airborne lidar data are becoming popular in Scandinavia to map postglacial faults and paleo-landslides many of which were recently discovered using these types of data (Malehmir et al., 2016; Mikko et al., 2015; Smith, Sundh, & Mikko, 2014; Sutinen, Hyvönena, & Kukkonen, 2014). Airborne lidar can also be used to create bathymetric models in shallow water (Doneus et al., 2015). Terrestrial applications of lidar can involve both stationary and mobile platforms.

Stationary terrestrial scanning is still the most common use of lidar as a survey method. A few examples include conventional topography, monitoring, cultural heritage documentation, and forensics (e.g., Vosselman & Maas, 2012). The 3D point clouds acquired from these types of scans can be matched with digital images taken of the scanned area to create realistic 3D looking models in a relatively short time in comparison to other technologies. Each point in the point cloud is given the color of the pixel from the image located at the same angle as the laser beam that created the point. Mobile lidar systems are used more and more frequently, e.g., for the monitoring along linear structures such as railroads, roads, and pipelines (e.g., Lato, Hutchinson, Diederichs, Ball, & Harrap, 2009; Leslar, Andelin, & Perry, 2009). Here, two or more scanners are attached to a moving vehicle to collect data along a path. These scanners are almost always paired with other kinds of equipment, including GNSS (global navigation satellite system) receivers and inertial measurement units.

2.8 Offshore Geophysical Measurements

Landslides, faults, and volcanoes are not restricted to terrestrial settings and many of the geophysical methods described above can also be applied offshore. Offshore landslides are a major concern for coastal communities and infrastructures as well as for those concerned with the safe development and exploitation of seafloor resources. Because of their unique geological and hydrologic settings, offshore landslides can be up to two orders of magnitude bigger in terms of mass flow than those that occur on land (SafeLand, 2010). Volcanic hazards and eruption histories of island arcs can be better studied at sea, thanks to volcanic deposits that are better preserved underwater than on the volcano's flanks above sea level. Major fault systems, such as the North Anatolian Fault (west of the destructive Izmit 1999 earthquake, Armijo et al., 2005) that constitute a potential hazard in the vicinity of Istanbul, continues from terrestrial observations to offshore where needs to be accurately mapped to establish the faults history and insights into the future. For completeness we list some specific marine geophysical methods in common use and their application basis in the following sections.

2.8.1 *Multibeam Systems*

Seabed topography is conventionally mapped using vessel-based acoustic echo-sounding devices, such as multibeam bathymetric systems. Through digital processing techniques and with the help of differential positioning systems (DGPS) these data provide precise shaded relief topographic maps. Such detailed maps are used for mapping landslide scars and deposits, volcanic deposits, and faults (e.g., Locat & Sanfaçon, 2000). Sub-bottom profiling, also called "sediment echo-sounding," can be used to identify and characterize sediments layers. This profiling method provides acoustical images of the seafloor for morphological mapping. Many sub-bottom profiling units can be towed behind or under a vessel or remotely operated with an autonomous underwater vehicle (AUV).

2.8.2 *Imagery*

Similar to aerial or satellite imagery, underwater photography is a powerful tool for morphological mapping. For example, scarps associated with recent earthquakes can be directly observed on the seafloor (Fig. 5, Armijo et al., 2005). This imagery provides direct observations to study fine-scale morphology and geology of those scarps, their distribution, and geometry. It can be carried out by towed camera systems (they usually require large support vessels and operate with limited survey speed) or with the use of unmanned remotely operated underwater vehicle (ROV).

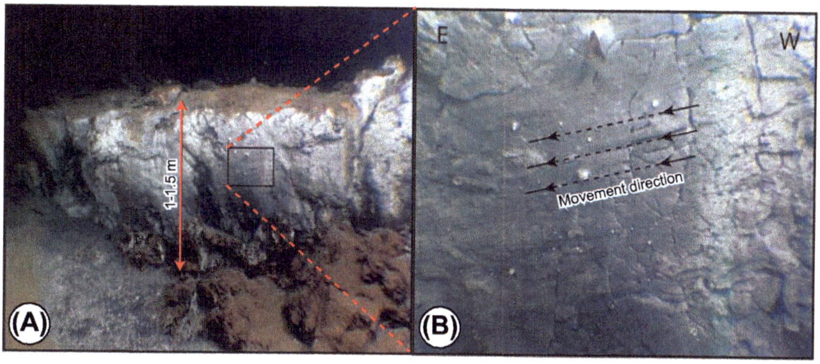

Figure 5 Sea-bottom pictures of earthquake scarp along the Marmara fault, west of Istanbul. (A) Vertical free face 1–1.5 m high corresponding to a recent earthquake rupture and (B) enlargement (*box* in A) showing slickensides with 15 degrees rake and 4–6 m net slip. *Figure from Armijo, R., Pondard, N., Meyer, B., Ucarkus, G., Mercier de Lépinay, B., & the MARMARASCARPS cruise party (2005). Submarine fault scarps in the Sea of Marmara pull-apart (North Anatolian Fault): Implications for seismic hazard in Istanbul.* Geochemistry, Geophysics, Geosystems, 6, *1–29.*

2.8.3 Active Offshore Seismics

Marine active-source refraction and reflection surveys have been used since the 1990s for fault detection and delineation (e.g., Laigle et al., 2008; Hornbach et al., 2011), characterization of submarine and coastal landslides and their related tsunamis potentials (e.g., Hornbach et al., 2011; Vanneste et al., 2015) as well as imaging volcanic structures at sea (e.g., Park, Morgan, Zelt, & Okubo, 2009). These methods were initially developed for and have heavily been exploited by the petroleum industry for which they typically penetrate several kilometers and have resolution in the order of tens of meters. High-resolution seismic data for marine geohazard purposes might penetrate several tens of meters and resolve features that are tens of centimeters thick (vertical resolution). Such data can be obtained using several source systems (e.g., Baldwin et al., 2004), such as water guns (20–500 Hz), airguns (100–1500 Hz), boomers (300–3000 Hz), sparkers (50–4000 Hz), and a variety of chirp systems (500–200,000 Hz).

3. EXAMPLES AND CASE STUDIES

3.1 Landslides

When studying landslides a few major factors should be considered such as hydrogeology at the site, strength of the materials against shearing and slide, and general slope around the site (e.g., Cruden & Varnes, 1996). Any or all of these factors can influence to varying degrees the initiation of a slide and each can be pivotal to the occurrence of one. Erosion is another factor that can be linked to one of the abovementioned factors. This is particularly important in areas close to rivers, lakes, and seas where tides and water level fluctuations contribute significantly to riverbank and shoreline erosion. Historical data and creeping of surface materials may also provide clues to susceptibility of an area to slide movement. There are cases where volcanic activities and heavy rainfall have led to huge landslides (e.g., Benito, Cepeda, & Martinez Diaz, 2004; Siebert, 2004) and in those cases knowledge of the thickness of the loose volcanic ashes and general slope in the area of investigation is important and must be carefully understood. It is well known that earthquakes can also trigger landslides.

Landslides are known to cause tsunamis. A classic example would be the tsunamis associated with the 1964 Alaska earthquake, which triggered localized subaerial and submarine landslides (e.g., Historic earthquakes, 1964). Up to five local slide-generated tsunamis were reported within minutes of the onset of shaking. There are a number of published accounts illustrating the usefulness of employing conventional geophysical methods for landslide studies, to understand areas prone to them and why they occur based on studying the aftermath of landslides (e.g., Bogoslovsky & Ogilvy, 1977; Jongmans & Garambois, 2007; Malehmir, Bastani et al., 2013; Malehmir, Saleem, et al., 2013; Méric et al., 2005; Socco, Jogmans, et al., 2010).

Given the abovementioned parameters, it is logical to employ resistivity-based methods (e.g., Lundström, Larsson, & Dahlin, 2009; Rankka et al., 2004) to study saturation condition, bedrock geometry, and depth and electrical properties of any overburden materials (for example, sand versus clay). RMT and EM-based methods have also successfully been used in areas close to rivers or lakes where traditional ERT-based methods suffer (e.g., Bogoslovsky & Ogilvy, 1977; Dahlin, Löfroth, Schälin, & Seur, 2013; Shan, Bastani, Malehmir, Persson, & Engdahl, 2014; Shan, Bastani, Malehmir, Persson, Lundberg, 2016).

Landslides often, but not always, occur in settings where there are steep slopes (gravity-related slide). There are particular and peculiar types of landslides known as quick-clay slides that occur primary in high-latitude countries including Sweden, Norway, Canada, Russia, and in the United States (in Alaska). Quick clays, originally deposited in a marine environment (about 10—12 kyr ago), lose part or all their salt content as a result of leaching by freshwater when they are uplifted above sea level following isostatic adjustment (Bjerrum, 1954). These types of landslides can occur at site with moderate slope, with little warning, and can be quite disastrous and fatal (Fig. 2).

Controlled samples of quick clay can liquefy with minimal stress changes (the touch of a human hand in the laboratory), while larger deposits in nature are less sensitive and require greater stress changes to mobilize, such as increased saturation by excess rainfall (e.g., L'Heureux, Hansen, Longva, Emdal, & Grande, 2010, 2012; Lundström et al., 2009; Vanneste et al., 2015). Examples of quick-clay slides from Sweden and Norway include Tuve 1977 (9 fatal causalities, 7 severed electric cables, and 65 diminished houses; Nadim, Pedersen, Schmidt-Thomé, Sigmundsson, & Engdahl, 2008), Vagnhärad 1997 (no human fatalities), which occurred along a clay slope, presumably caused by elevated stress on the slope, likely triggered by heavy rainfall (Löfroth & Kjellberg, 2003). The 2006 Munkedal quick-clay landslide in Sweden occurred in a reasonably high traffic highway, which was speculated due to construction activities combined with heavy rainfall (Nadim et al., 2008). Because of the experience of several authors of this paper with quick-clay landslides, a special focus has been given to them here. Several examples are provided that illustrate the use of various near-surface geophysical methods to characterize this type of landslide, but any kind of landslide can benefit from using similar approaches.

The likelihood for movement of a quick-clay slide depends on the existence of leached marine clay, topographic relief, and the potential of changes to slope stability due to erosion or excavation. Knowledge about thickness of the clay materials, degree of internal heterogeneity, bedrock depth, and slope are important to characterization of quick-clay landslides. Thickness and extent of quick-clay layers or lenses determine the run off distance in the case of slide movement. While many geophysical methods can illuminate the depositional structure and bedrock topography, resistivity methods have proven to be particularly powerful for delineating quick-clay deposits because of the contrast in salinity to surrounding unleached, often marine, clays.

A group of institutions and researchers in Norway has recently completed a large-scale study on advanced identification of quick clay, using both geotechnical and geophysical methods (NIFS, 2015). Quick clay is definitively identified in soil samples based on sensitivity and remolded shear strength thresholds. Indirect correlation between resistivity and remolded shear strength from measurements at several sites in Norway is convincing

(Long et al., 2012) (Fig. 6). This correlation has been generalized by NIFS (2015). Joint geophysical and geotechnical quick-clay delineation, based on resistivity measurements in 1D (R-CPTu), 2D (ERT), and 3D (AEM) represents an efficient grouping of geophysical methods for characterizing this particular geohazard. Numerous case studies illustrate the capability to delineate quick clay from unleached clay (Anschütz et al., 2016; Bazin & Pfaffhuber, 2013; Donohue et al., 2012; Pfaffhuber, Bazin, & Helle, 2013; Rømoen, Pfaffhuber, Karlsrud, & Helle, 2010; Solberg et al., 2008). Within the scope of this paper, we have focused on a complex case where a multi-geophysical approach sheds light on the structure of a complicated quick-clay deposit in Sweden.

In Sweden where quick-clay landslides are part of the historical record and a recognized threat, an extensive geophysical program supported by the GWB program of the Society of Exploration Geophysicists (SEG) and

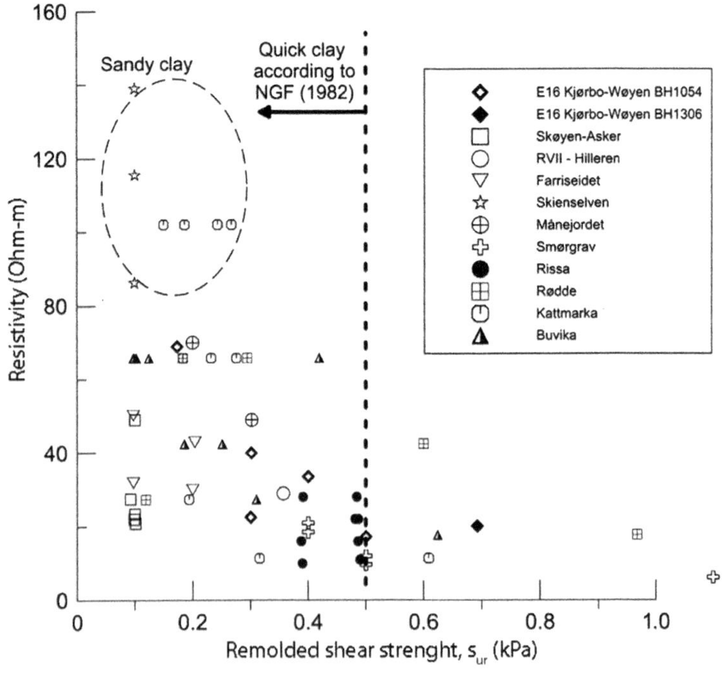

Figure 6 Resistivity and remolded shear strength measured at some Norwegian quick-clay sites. *Modified from Long, M., Donohue, S., L'Heureux, J.-S., Solberg, I.-L., Rønning, J.S., Limacher, R., …, Lecomte, I. (2012). Relationship between electrical resistivity and basic geotechnical parameters for marine clays.* Canadian Geotechnical Journal, 49*(10), 1158–1168.*

various other organizations used various surface and more recently airborne geophysical methods to understand the subsurface geology that has contributed to the generation of landslides at the site (Malehmir, Bastani et al., 2013). Given the conductive nature of the marine clays combined with 100% saturation at just 1 m below ground surface, a full complement of geophysical methods was evaluated at this site, which in addition to ERT, RMT, GPR, and refraction seismic methods, included reflection-type seismic surveys. P-wave reflection seismic surveying was employed at this site using both 2D (e.g., Malehmir, Saleem, et al., 2013) and 3D (Lundberg, Malehmir, Juhlin, Bastani, & Andersson, 2014) approaches and at a few localities SH—SH reflection surveys (Krawczyk, Polom, Malehmir, et al., 2013) were also performed.

A known retrogressive-type quick-clay landslide scar near the Göta River site (source of drinking water for more than 700,000 people) was imaged on vertical component (P-wave) reflection sections (Fig. 7). Reflection seismic data at this site were obtained using 120 receivers spaced 4 m apart and a combination of an accelerated weight drop and 20—40 g of

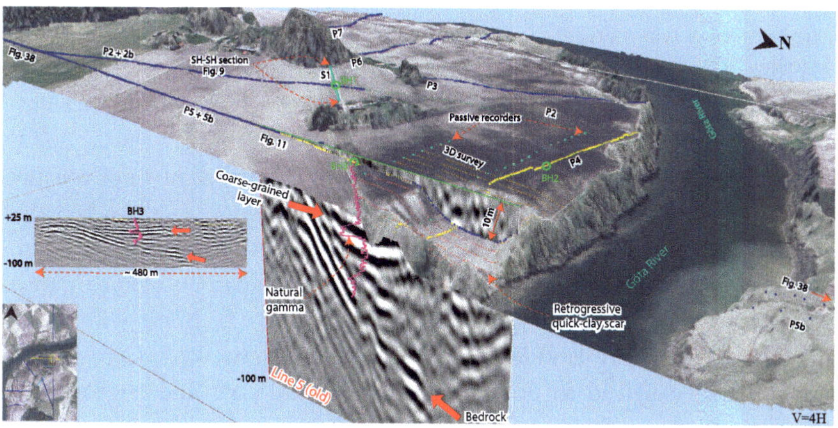

Figure 7 Example of high-resolution reflection section from a quick-clay landslide site in southwest Sweden by the Göta River showing two sets of reflectors (see the *red arrows*), one associated with a coarse-grained (mainly sandy) layer (natural gamma drop) and another one with bedrock. Quick clay in the site is observed often above the coarse-grained layer implying its importance when studying quick clays in the site. The scar appears to be close to the top of the coarse-grained layer suggesting this to be likely a slip surface. *Based on Malehmir, A., Saleem, M.U., & Bastani, M. (2013). High-resolution reflection seismic investigations of quick-clay and associated formations at a landslide scar in southwest Sweden.* Journal of Applied Geophysics, 92, 84—102.

explosives to generate seismic waves (Malehmir, Saleem, et al., 2013). A conventional reflection data processing workflow was followed and, occasionally, refraction static corrections were applied to account for the extreme low velocity (dry curst) near the surface.

Two sets of reflections were observed at this site with uniquely different characteristics. A horizontal reflection at about 20 m depth onlapping a moderately dipping reflection that comes very near to the ground surface but does not outcrop. The horizontal reflection was interpreted to be the top of coarse-grained (sand and gravel) materials (Lundeberg et al., 2014; Malehmir, Bastani et al., 2013; Malehmir, Saleem, et al., 2013), and the dipping reflection was identified as the top of bedrock. This interpretation was later confirmed using borehole sediment sampling (direct observations) and downhole physical property measurements (Salas-Romero et al., 2016). Natural gamma plot clearly shows a drop associated with the coarse-grained materials at about 20 m depth (Fig. 7). As evident from the natural gamma log, the vertical component data had a maximum vertical resolution of about 5 m since they could not resolve top and bottom of the coarse-grained layer. Note that seismic reflection detection potential (usually one thirtieth of a seismic wavelength) is usually much better than the vertical bed resolution (a quarter of a wavelength).

None of the other geophysical methods employed at the site were able to detect this layer within the unlithified sediments above the bedrock. Refraction (Adamczyk et al., 2013, 2014), ERT, and RMT (Shan et al., 2014, 2016) methods were helpful to partly resolve the bedrock but not reliably delineate the sandy layer within the clayey sediments. The sandy layer was also clearly observed in the 3D reflection data acquired at the site and proved to be a large-scale feature and not limited to a particular line or area.

The significance of the sandy layer and shallow bedrock within the context of quick-clay landslides should not be overlooked. One of the scenarios that can lead to the formation of quick clays and trigger their movement is the infiltration of surface water along the bedrock surface at shallow depths with the channeling of that water into the sand layer until it effectively becomes a confined aquifer where it leaches salt from the overlying clays (or in combination with surface infiltration). The coarse-grained layer can also act as a slip surface, particularly if it occurs closer to the surface (Fig. 7).

In the Tuve landslide, people visiting the site after the slide noticed an unusually large amount of sand on the surface (Wang et al., 2016). This supports the notion that the coarse-grained materials often act as a slip surface for a number of quick-clay slides in Sweden. For the Tuve landslide, the increase in pore-pressure combined with erosion of the layer near the riverbank may have been a triggering mechanism (e.g., Löfroth, Suer, Dahlin, Leroux, & Schälin, 2011).

At the same site, shear wave data were acquired using a 120 SH-geophone (transverse component) land streamer with 1 m sensor spacing and a microvibrator (ELVIS) seismic source with source points every 2–4 m (Fig. 8). High-quality SH-wave data were acquired in an attempt to resolve the depth gap between the P-wave data and the electrical and surface-wave-based methods (Malehmir, Bastani et al., 2013; Malehmir,

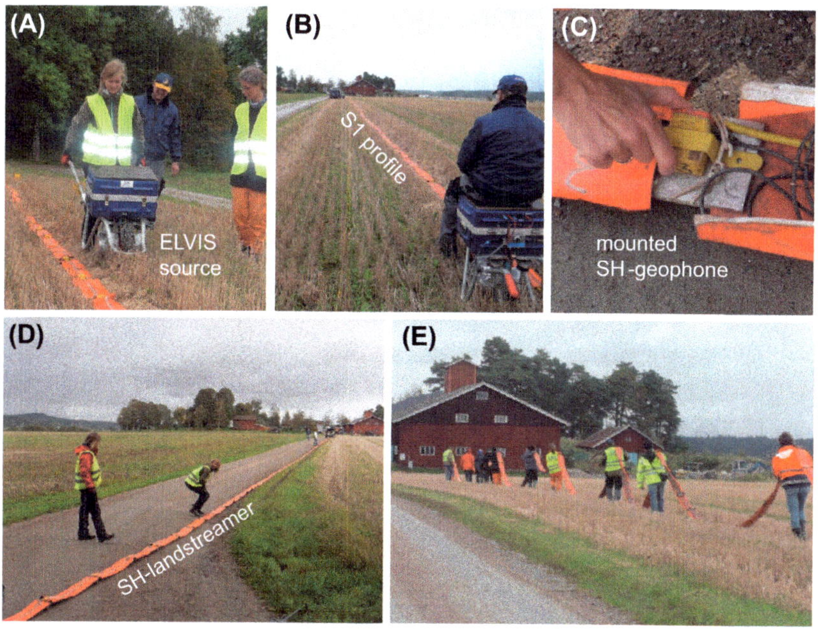

Figure 8 A selection of photos from the shear-wave (SH–SH) reflection seismic fieldwork at the Swedish quick-clay site using the equipment developed by LIAG (Krawczyk, Polom, & Beilecke, 2013; Krawczyk, Polom, Malehmir et al., 2013; Polom, Druivenga, Grossmann, Grüneberg, & Rode, 2011). (A) ELVIS microvibrator moved between two shot points, (B) layout on the stubble field with the land streamer (*orange*) and seismic source, (C) a close view of the transverse component geophone mounted on the streamer (1 m spacing), (D) layout on the road (see S1 in Fig. 7), and (E) manual land streamer shift on the stubble field. *Modified from Krawczyk, C.M., Polom, U., Malehmir, A., & Bastani, M. (2013). Quick-clay landslides in Sweden – insights from shear-wave reflection seismics and geotechnical integration.* Proceedings of the near surface geoscience 2013, *Bochum/Germany, 5 pp. http://dx.doi.org/10.3997/2214-4609.20131348. Data along the gravel road are presented here.*

Saleem, et al., 2013; Sauvin G., personal communication, 2011) and that possessed lower resolution along the very undulating bedrock surface. Roll-along survey geometry was used for the data acquisition (Krawczyk, Polom, & Beilecke, 2013; Krawczyk, Polom, Malehmir, et al., 2013).

Given the high quality of these SH–SH data, conventional processing was employed up to the stacked section where an explicit steep-dip finite difference time migration was applied (e.g., Yilmaz, 1987) to correct for the steep flanks of the bedrock (Polom et al., 2013). Love waves were remarkably weak, and reflection hyperbolas could easily be identified without specific prestack data enhancement algorithms. This is the advantage of surveying along a gravel road above natural unlithified sediments, which, in this case represents a strong decrease in shear-wave velocity between the high velocities within the roadbed to very low velocities with the natural layers below.

Reflections from near the ground surface down to about 50-m deep can be interpreted on the SH–SH stacked reflection section with unprecedented high resolution (Fig. 9). None of the other methods employed at

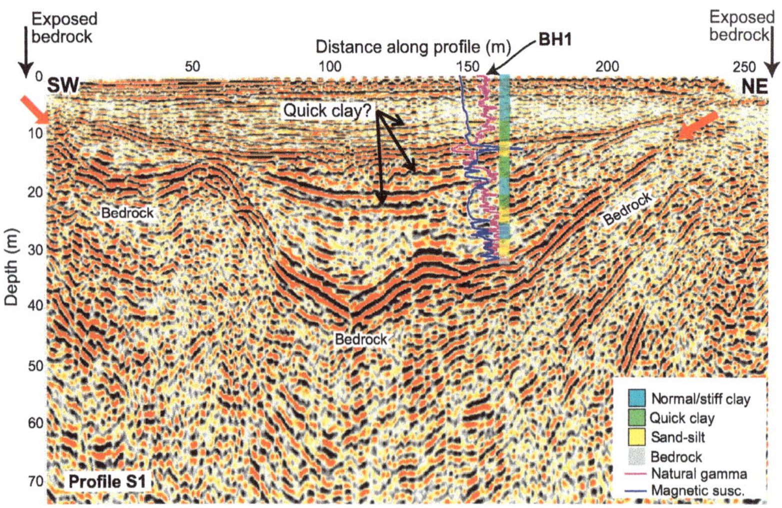

Figure 9 SH–SH reflection section (see S1 in Fig. 7) along the gravel road (see Fig. 8D) showing two distinct reflectors associated with coarse-grained materials, above which quick clays were observed, and an undulating one associated with bedrock. The coarse-grained materials show distinct natural gamma decrease and magnetic susceptibility increase (Salas-Romero et al., 2016). *Modified from Polom, U., Bagge, M., Wadas, S., Winsemann, J., Brandes, C., & Krawczyk, C.M. (2013). Surveying near-surface depocentres by means of shear wave seismics.* First Break, 31, *63–75. http://www.earthdoc.org.*

the site were capable of resolving such fine structures within the normally consolidated glacial sediments above the bedrock. Horizontally layered sediments are visible in the upper 40 m above the top of basement varying vertically between c. 10–40 m depth and outcropping on both ends of the line (see Fig. 7). The top of basement has a bowl-shaped morphology with the center of this synclinal structure near the middle of the line. Several distinct sediment sequences are separated by high-amplitude unconformities (Fig. 9) that were later calibrated using borehole data and downhole logging measurements (Salas-Romero et al., 2016).

The two most prominent reflections are associated with coarse-grained materials that exhibit clear decreases in natural gamma counts and increased magnetic susceptibility (Fig. 9). They have similar characteristics to the one observed on P-wave data (Fig. 7) and other parts of the site (Malehmir, Saleem, et al., 2013). Shear-wave velocities calculated during reflection data processing are extremely low with values of 100–120 m/s. This slow velocity suggests that materials at the site have low shear modulus and will not be conducive to construction in this area, at least from a geotechnical point of view. Given the bowl-shaped geometry of the bedrock, we do not think this region is prone to natural landslides, but landslides triggered by human activities are possible (e.g., constructions and unloading the materials by trenching). The shear-wave data from this site possess a vertical resolution in the order 1 m when calibrated with the natural gamma data.

Geophysical surveys have also been used during reconstruction. Some examples include Banda Aceh after the devastating tsunami (Polom, Arsyad, & Kümpel, 2008) and larger-scale landslide studies have been undertaken in Nordic countries (Polom et al., 2010; Polom and Krawczyk, 2014). The area around Trondheim bay, central Norway, for example, occasionally suffers from both onshore and offshore landslides. The urban growth in the harbor requires determination of in situ soil conditions before construction and safe expansion of the infrastructure.

To delineate subsurface structures and better understand any potential geohazard, an SH–SH reflection seismic survey, similar to the one previously described in Sweden, but using a stronger seismic source was carried out (Polom et al., 2010) to complement a dense network of high-resolution single-channel marine seismic profiles over the deltaic sediments in the Trondheim fjord (L'Heureux et al., 2010). The onshore profile (SH–SH) (Fig. 10) was acquired along roads and parking places. Seismic images of the sediments, overlying the bedrock at about 200–250 m depth, exhibits distinct sequences that can be tied seismo-stratigraphically to the

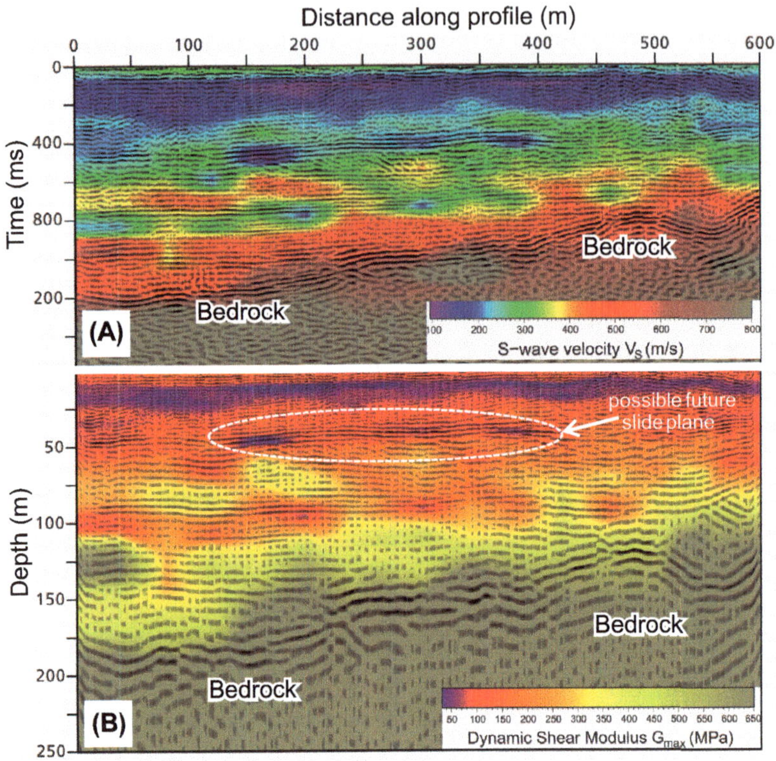

Figure 10 SH–SH reflection section from Trondheim harbor. (A) Poststack time-migrated section with superimposed shear-wave interval velocity (V_S) used for normal move-out (NMO) corrections. (B) Poststack depth-migrated section with dynamic shear modulus (Gmax) estimated using the NMO velocities. Gmax was derived using a constant density of 1900 kg/m^3. *Modified from Polom, U., Hansen, L., Sauvin, G., L'Heureux, J.-S., Lecomte, I., Krawczyk, C. M., … Longva, O. (2010). High-resolution SH-wave seismic reflection for characterization of onshore ground conditions in the Trondheim harbor, central Norway. In R. D. Miller, J. D. Bradford & K. Holliger (Eds.),* Advances in near-surface seismology and ground-penetrating radar, SEG, Tulsa, *297–312. ISBN:978-1-56080-224-2.*

well-calibrated marine sequences (Hansen et al., 2013). Important to note here is that distinct, pure shear reflection events enabled detailed shear-wave velocity analysis down to the bedrock surface (Fig. 10) allowing soil stiffness to be roughly obtained from the interval velocities by assuming a constant density of 1900 kg/m^3 (Fig. 10B).

Shear-wave velocities of about 100 m/s occurring in the upper 50 m of the fjord–deltaic sediment succession suggest low sediment stiffness (i.e., 50–100 MPa) directly below the 10–15 m thick and stiffer artificial fill.

In addition, a possible future slip plane can be interpreted below this layer. Even though not continuous, lowered shear modulus values occur piecewise along a strong and continuous reflector at about 50 m depth (Fig. 10B). These results indicate that the SH-wave seismic reflection method is well suited for urban geohazard investigations.

Higher data quality could have been obtained if night operation was undertaken; the resulting interval shear-wave velocities can provide proxy in situ soil stiffness (at small strain level) and be used as a first order estimate for geotechnical characterization. There are no borehole data to confirm the survey resolution but according to the shear-wave velocities used for imaging and seismic frequencies generated (maximum 100 Hz), we propose a vertical resolution on the order of 1–5 m down to about 200 m depth. The penetration depth is at least 200 m after which there are no clear reflections to confirm if deeper penetration was possible. No amplitude decay curve analysis on shot gathers is available to elaborate on the penetration depth.

With respect to the use of electrical and EM methods, Shan et al. (2014) provided a detailed account on the RMT and ERT measurements along four lines from the site. According to Solberg (2007) the marine clays with the salt content below 5 g/l in their pore water can be considered as either sensitive or quick. This was used as a criterion to select both ERT and RMT methods for delineating quick-clay materials in the site. The RMT data were acquired using EnviroMT instrument (Bastani, 2001) with 10 m station spacing along the lines. The EM noise level in the area was reasonably low resulting in good quality RMT data (Shan et al., 2014). A multielectrode system (ABEM Terrameter LS) with a Wenner electrode-array configuration and minimum electrode spacing of 5 m were used to collect ERT data along the RMT lines.

Individual and joint 2D inversions were carried out to model the electrical resistivity of subsurface materials that control the formation of quick clays and also occurrence of the landslides in the area. The RMT data used in the inversions were determinant with apparent resistivities and phases estimated at nine target frequencies in the one octave frequency bandwidths. Details of the algorithms used for the inversions of ERT and RMT data can be found in Shan et al. (2014). RES2DINV (Loke, 2001) was used for the single inversion of the ERT data and EMILIA (Kalscheuer, Garcia, Meqbel, & Pedersen, 2010), which is an extended version of the REBOCC program (Siripunvaraporn & Egbert, 2000) for the 2D single and joint inversions of the RMT and ERT data.

Resistivity models were produced from 2D inversions of RMT and joint ERT and RMT data (Fig. 11) along the seismic profile (Fig. 7). Bedrock is partly resolved on these data. A high-resistivity zone above the bedrock corresponds with the location of the reflection correlated with the

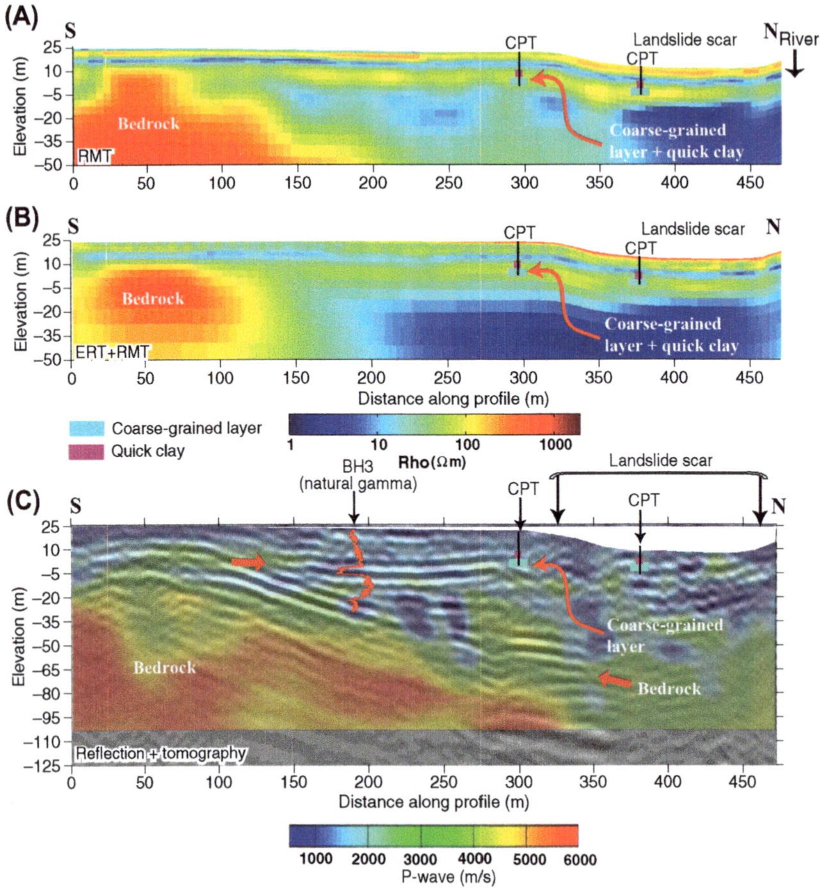

Figure 11 Resistivity and refraction tomography models along profile 5 (for location see P5 in Fig. 7) from (A) single inversion of the radio magnetotelluric (RMT) data, (B) joint inversion of electrical resistivity tomography (ERT) and RMT data, and (C) full-waveform refraction tomography (Adamzcyk et al., 2013, 2014) superimposed on the reflection section. Note various features identified in various methods used at the site. The coarse-grained layer and immediate materials above it show high-resistivity, which may imply leaching or freshwater flow in the zone. *ERT and RMT models are based on Shan, C., Bastani, M., Malehmir, A., Persson, L., & Engdahl, M. (2014). Integrated 2D modeling and interpretation of geophysical and geotechnical data to delineate quick clays at a landslide site in southwest Sweden.* Geophysics, 79, *EN61–EN75.*

coarse-grained layer (as well as the natural gamma drop), which suggests either there is freshwater flow within the layer or the freshwater flow is combined with leached materials above the layer (indicating presence of quick clay). If this is true, both of these methods have better vertical resolution (top and bottom) to resolve this shallower resistive layer than the seismic refraction and reflection methods, but lack imaging potential for the deeper bedrock because of the thick and low-resistivity marine clays.

The application of satellite-based InSAR is useful to estimate the extent of a landslide as well as the magnitude of its movement, especially when the slide area is large. GPS measurements and other positioning measurement techniques can be used for this purpose too, but only at a limited number of discrete measurement points and less cost-effective. Radar scenes provide a good overview of the areas with movement, and can provide

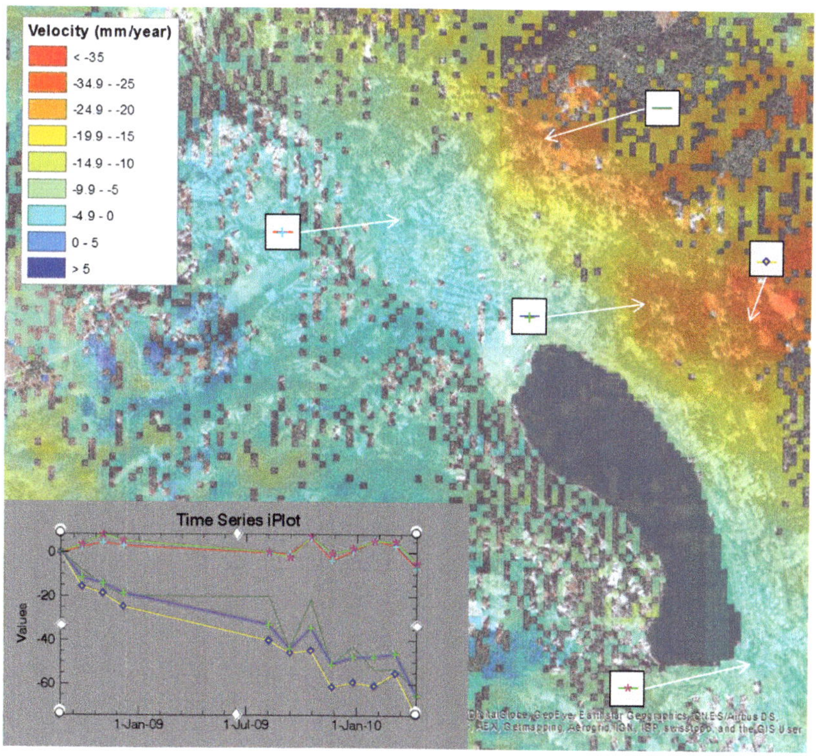

Figure 12 InSAR-based terrain deformation analysis at Nainital Lake, India. Mean deformation rate in mm/year based on an EnviSAT image stack. The inset figure shows a velocity time series from August 2008 to August 2010 (in mm/year). *InSAR*, interferometric synthetic aperture radar.

time-lapse measurements over many years, provided that historical data are available. By using data from a sensor with a short wavelength and by using 25 or more scenes, one can obtain a long and precise time series of the region's movement in two dimensions. A map example of InSAR deformation analysis around Nainital Lake in India (Vöge et al., 2015) shows a good number of coherent pixels around the lake (Fig. 12). Most of the coherent areas show movement with a mean velocity of −2.5 to +1.5 mm/year and, thus, considered stable. Along the mountain slopes northeast of the lake, however, the measurements identify a large area of deformation with rates that reach up to over −35 mm/year (movement away from the sensor that is downslope in this case).

3.2 Flood-Damage Assessments

Flooding can cause severe erosion and damage infrastructures. Flooding combined with landslides can become extremely dangerous with floodwaters carrying large amount of debris (clay and large boulders) from slides in highland areas. The 2011 devastating Rio de Janeiro flood became disastrous as a result of the many landslides and debris flow (Fig. 3A). Geophysical methods can map the extent and thickness of debris flow (airborne methods if large areas and ground-based for specific and local scale cases) as well as assess damaged infrastructures within the flood areas.

In response to the damaging flood in Serbia, and, Bosnia and Herzegovina in 2014, a call for assistant was made (Komatina, Urosevic, & Suto, 2014) and another project supported by the Geoscientists Without Borders of the Society of Exploration Geophysicists was started with the aim of illustrating the usefulness of near-surface geophysical methods to address some of the issues related to floods. Seismic refraction and reflection, multichannel analysis of surface-waves (MASW; Park et al., 1999), and electric resistivity methods with digital terrain models using drone surveys were performed (Komatina, Suto, & Urosevic, 2015). Urosevic et al. (2016) summarized that there were good agreements between seismic and electric images of the subsurface materials in their project.

Suto (2016) focused on MASW aspect of the project with comparison of MASW and resistivity surveys presented in Suto, Arsenovic, Sugawara, and Aizawa (2016). Correlation between shear-wave velocity data derived from MASW and ERT results from a landslide-prone slope greatly improved confidence and resolution of results (Fig. 13). The transitions from blue to

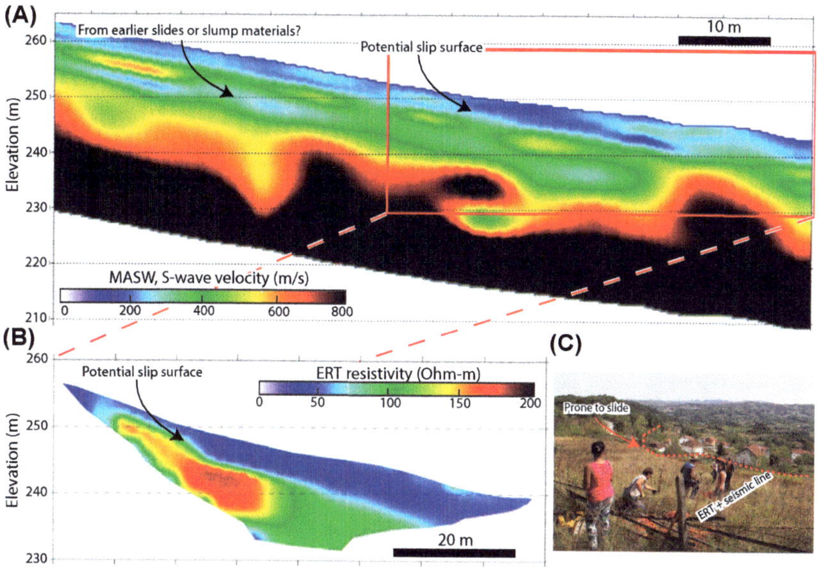

Figure 13 A comparison between (A) S-wave velocities derived from multichannel analysis of surface-waves (MASW) and (B) resistivity sections from (C) an area prone to slide in Bosnia and Herzegovina. Both geophysical models suggest a potential slip surface and large pockets of sediments in the lower end of the hill. *Based on Suto, K., Arsenovic, S., Sugawara, J. & Aizawa, T. (2016). An investigation of a landslide-prone area by MASW and resistivity surveys, in Proceedings of Symposium on the Application of Geophysics to Engineering & Environmental Problems (SAGEEP) 2016, Denver, USA and published with the permission of the Environmental and Engineering Geophysical Society.*

green in both sections are interpreted as a potential slip surface. The S-wave section includes a low-velocity layer under the high-velocity layer. These layers may indicate debris from previous landslide episodes.

Suto and Kristinof (2014) reported a successful application of MASW survey on a flood-damaged mountain road in Australia, where geotechnical drilling was risky due to the extent of the damage (Fig. 14A). Resulting S-wave velocity sections helped identify potential places where the materials, mainly clay, were washed away requiring geotechnical grouting or similar approaches for its reinforcement (Fig. 14B).

Shear-wave velocity models can provide an estimate of dynamic elastic properties, using various assumptions, and help to design appropriate Gabion walls. An example reinforcement design and how it is constructed in practice provides valuable insights into the value of shear-wave velocity measurements (Fig. 15).

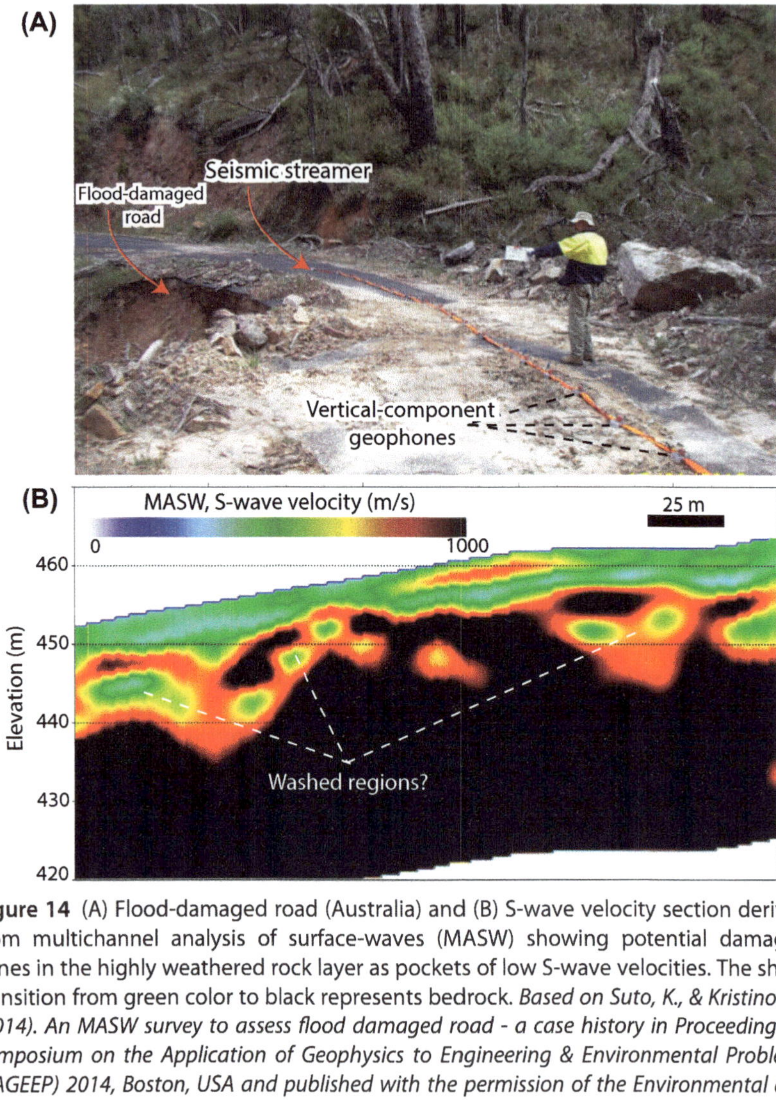

Figure 14 (A) Flood-damaged road (Australia) and (B) S-wave velocity section derived from multichannel analysis of surface-waves (MASW) showing potential damaged zones in the highly weathered rock layer as pockets of low S-wave velocities. The sharp transition from green color to black represents bedrock. *Based on Suto, K., & Kristinof, R. (2014). An MASW survey to assess flood damaged road - a case history in Proceedings of Symposium on the Application of Geophysics to Engineering & Environmental Problems (SAGEEP) 2014, Boston, USA and published with the permission of the Environmental and Engineering Geophysical Society.*

(A)

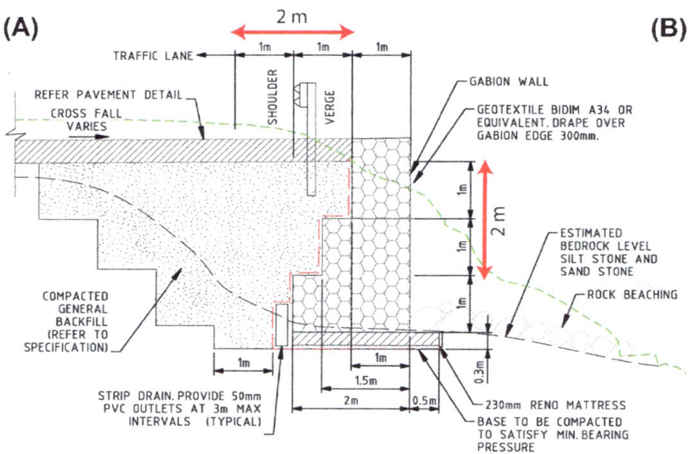

(B)

Figure 15 (A) Design of Gabion wall and (B) an example of it for a flood-damaged road. *Based on Suto, K., & Kristinof, R. (2014). An MASW survey to assess flood damaged road - a case history in Proceedings of Symposium on the Application of Geophysics to Engineering & Environmental Problems (SAGEEP) 2014, Boston, USA and published with the permission of the Environmental and Engineering Geophysical Society.*

3.3 Rockfalls

Rockslides and falls may occur in mountainous areas and fjords but also in low lands when deep construction cuts expose unstable rock blocks. Numerous mechanisms can lead to a rockfall such as rotation, toppling, and sliding (Cruden & Varnes, 1996). The driving mechanism is predominantly a combination of fractures often filled with soft material to "lubricate" the slide and extensive pore water pressure destabilizing a block of rock. These fractures and sliding planes can be targeted by geophysical surveys. Rockfalls have caused tsunamis. One well-known example is from Alaska again associated with a magnitude 7.9 earthquake. A rockfall following that earthquake generated a wave (from the rockfall into fjord) that traveled more than 500 m onshore with an estimated 30 m high wave, sinking fishing boats and killing two people (Miller, 1960).

Various satellites and ground-based remote sensing methods can be applied to assess whether an unstable rock block is moving and potentially failing (Bardi et al., 2016). For example, a large potential rockslide in Norway, Åknes (Blikra, 2012), has been studied using various geophysical (Heincke et al., 2010) and remote sensing methods (Bardi et al., 2016). As a result of these surveys an operational monitoring and warning system has been designed and is currently being used. In a worst-case scenario, Åknes failure may result in a tsunami engulfing several communities in close proximity (Eidsvig et al., 2009). The monitoring system aims to notify emergency responders of resident in threatened areas so action can be taken days prior to its potential failure.

As an example, we present results from a rockslide study in western Norway that included AEM, ERT, satellite-based InSAR, ground-based InSAR, and LiDAR surveys. The Flåm fjord (Fig. 16) is among Norway's most famous tourist destinations with up to 450,000 visitors and more than 150 cruise ships a year arriving in the area. Flåm also hosts the northern end of the historic Flåm railway located along one of the steepest rock faces in the world that allow train passes. More than 700,000 passengers use the railroad every summer for transportation and for viewing countryside. The main road between Bergen and Oslo (E16) passes through Flåm, and bypasses the fjord. Historical evidences of large rockslides have been documented in the area with present-day ground movement also evident (Blikra, 2012). The areas that are subject to potential rockslides experience creeping rock and debris masses (Fig. 16).

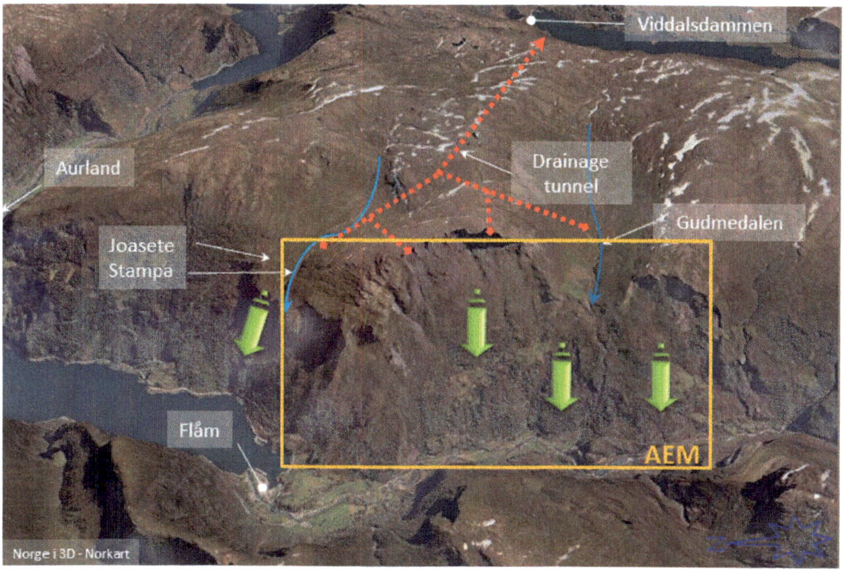

Figure 16 Aerial photo draped over topographic model *(courtesy of www.norgei3d.no.)* showing the Flåm valley and fjord, and known rockslides and creeping movements (*green arrows*) of both massive rock (fjord) and loose debris (valley) partly driven by pore water delivered by the Stampa and Gudmedalen catchments. *Red lines* indicate a potential water drainage tunnel system. *Modified from Pfaffhuber, A.A., Grimstad, E., Domaas, U., Auken, E., Halkjær, M., & Foged, N. (2010). Airborne EM mapping of rock slides and tunneling hazards.* The Leading Edge, *29, 956–959.*

The area most susceptible to and with the greatest potential rockslide was surveyed using AEM methods in 2009 (Pfaffhuber et al., 2010) with follow-up ground measurements focusing on a potential sliding planar structure identified with AEM data (Pfaffhuber et al., 2011). Complementing these results, detailed structure maps of the unstable rock units were generated with lidar scans. Furthermore, movement of the blocks facing the fjord was quantified with ground-based InSAR measurements. These data were integrated into a conceptual model of the slide (Blikra et al., 2012).

The success of the original study prompted an extension of the AEM survey into the lower Flåm valley to map debris from past slides that are creeping downslope and represent a hazard for the road and railway in the valley (Anschütz, Pfaffhuber, Domaas, Rosenvold, & St, 2015). The resistivity models derived from the AEM survey (Figs. 17 and 18) clearly highlight a conductive layer that we interpret as the base of debris and assumed to be

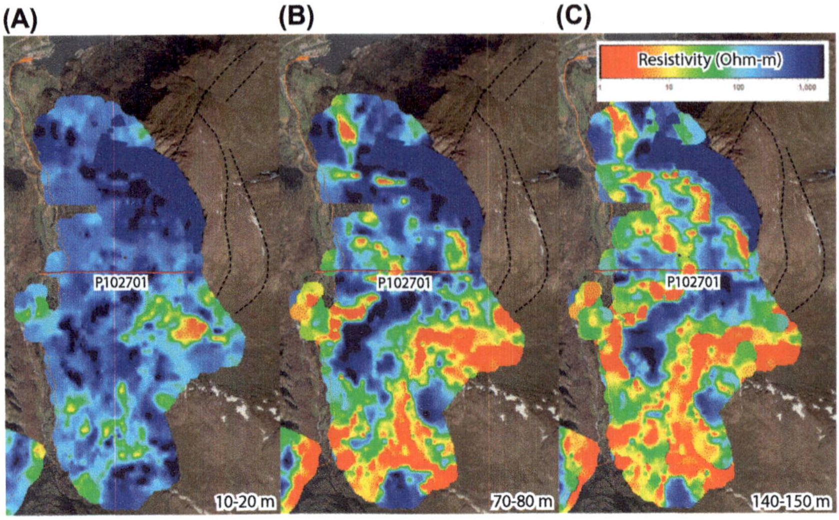

Figure 17 Resistivity depth slices for three representative depth intervals, (A) 10–20 m depth, (B) 70–80 m depth, and (C) 140–150 m depth, respectively, in the Flåm valley. Cross section shown in Fig. 18 is indicated as a *red line* here. *Modified from Anschütz, H., Pfaffhuber, A.A., Domaas, U., & Rosenvold, B. St. (2015). Combined airborne and ground geophysics as a first phase towards a landslide warning system – a Norwegian case study.* Österreichische Ingenieur- und Architekten-Zeitschrift, *160. (Jg).*

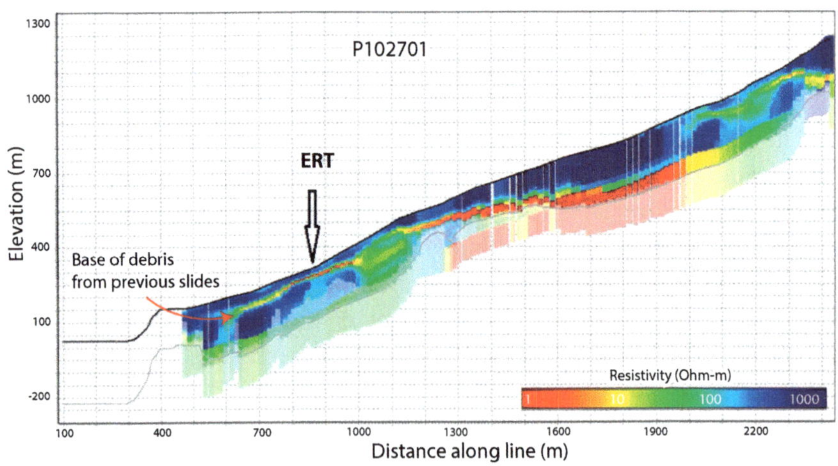

Figure 18 Section through the 3D resistivity model at profile 102701 (see Fig. 17). A thin, conductive layer is evident throughout the lower part of the slope. It is interpreted as the base of debris from past slides. The *gray line* indicates the individual penetration depth and thus the parts of the model (grayed out) with rather high uncertainty. *ERT*, electrical resistivity tomography. *Modified from Anschütz, H., Pfaffhuber, A.A., Domaas, U., & Rosenvold, B. St. (2015). Combined airborne and ground geophysics as a first phase towards a landslide warning system – a Norwegian case study.* Österreichische Ingenieur- und Architekten-Zeitschrift, *160. (Jg).*

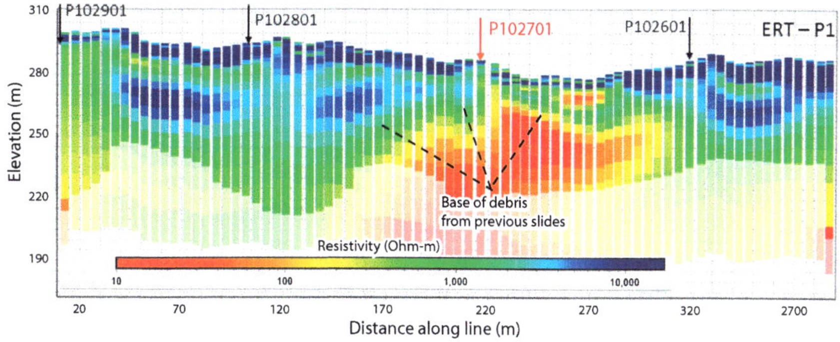

Figure 19 Electrical resistivity tomography (ERT) profile perpendicular to the airborne electromagnetic (AEM) flight lines and displayed from south to north. A shallow conductive anomaly is in agreement with the thin conductive layer imaged by the AEM survey. Note that the resistivity color scale is comparable to the AEM color scale but adapted to account for the larger resistivity aperture of ERT. *Modified from Anschütz, H., Pfaffhuber, A.A., Domaas, U., & Rosenvold, B. St. (2015). Combined airborne and ground geophysics as a first phase towards a landslide warning system — a Norwegian case study.* Österreichische Ingenieur- und Architekten-Zeitschrift*, 160. (Jg).*

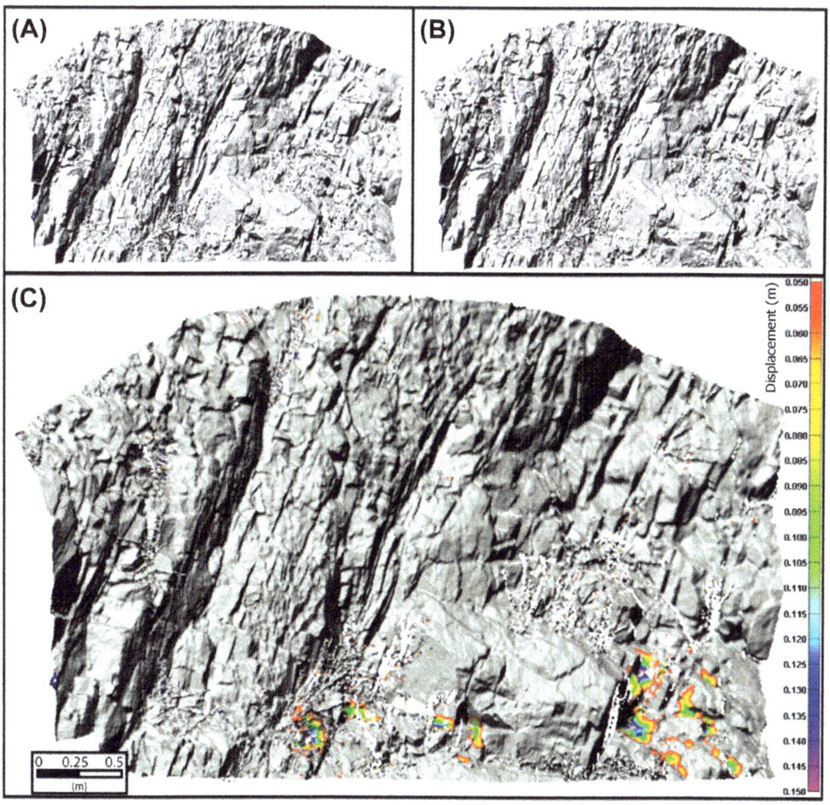

Figure 20 Volumetric change detection in a rock section near Kingston, Ontario (Canada), by comparison of successive scans in time (A) and (B) from a similar location. Color-coded zones in (C) represents between 5 and 10 cm of movement or release of rock blocks. Fig. 15 in Lato et al. (2009), licensed under CC BY 3.0.

rich in fine-grained water-saturated materials, and thus having low electrical resistivity.

Two additional ERT profiles (Fig. 19) were acquired and combined with drill investigations showed artesian water at the depth predicted from the airborne resistivity model (Anschütz et al., 2016). Preliminary monitoring from the well data shows a good correlation between precipitation and pore pressure in the interpreted slide plane. The first step for a monitoring and early warning system (locate the slide plane and borehole confirmation of a relationship between precipitation and borehole fluid pressure) was thus successfully achieved. In terms of resolution, ERT and AEM data show a vertical resolution of less than 5 m at about 20 m depth.

Repetitive lidar surveys can also be used for rapid appraisal of a potential rockfall site (Fig. 20).

Another option for rapid or real-time time-lapse monitoring of potential rockslide or rockfall areas is with ground-based radar systems (e.g., Ferrigno, Gigli, Fanti, & Casagli, 2015; Kristensen, Rivolta, Dehls, & Blikra, 2013) (Fig. 21).

Figure 21 Ground-based interferometry results (change detection between November 2015 and April 2016) draped over a 3D surface model derived from RPAS captured imagery. Green areas are considered stable. The *yellow line* outlines areas with block accumulation (blue) and block evacuation (red to yellow) in the talus cone below the rock wall.

3.4 Sinkholes and Subsidence

Sinkholes are generally circular to ellipsoidal surface depressions or collapse structures, caused by underground dissolution and subrosion processes. Diameters may range from a few meters to several hundred meters for large sinkholes (Fig. 22), thus posing a severe hazard for infrastructure and inhabitants.

Depending on the subsoil properties, strength of rock between the top of bedrock and the soluble rock, and the dissolution process, sinkholes may form gradually (e.g., Hamburg in Germany; Reuther et al., 2007) or collapse catastrophically and without warning forming steep-sided depressions. Regardless of the subsidence rate, these surface depressions can possess diameters and depths well in excess of a hundred meters with the soluble rock responsible hundreds of meters below the ground surface (e.g., Kansas in

Figure 22 Sinkhole, which formed catastrophically in central Kansas, United States, in an industrial area in the city of Hutchinson. It grew to within a few meters of the railroad tracks in the background before it was hauled with slope stabilizing grout. Surface seismic surveys were used to guide the grout placement that eventually arrested the migration of the sinkhole toward the railroad tracks. This sinkhole was the result of solution mining in the early 1900s. The sinkhole was first noticed by a conductor on one of the dozens of trains (freight and passenger) that use the tracks in the background. At one point, growth of the sinkhole was measured in centimeters per hour. *Photo by Richard D. Miller.*

the United States; Miller, 2003; Miller, Routh, et al., 2008; Miller, Villella, et al., 2008; Miller, Xia, Steeples, 2009; Münsterdorf in Germany, Gebregziabher, 2011; in Italy, Parise & Lollino, 2011). More examples on specific sinkhole types and formation processes are discussed in Waltham, Bell, and Culshaw (2005). Both types of sinkholes may cause severe damage to infrastructure, for example, c. 93 million dollars paid by insurance companies in Florida in 2009 (The Florida Senate, 2010); 38 million dollars loss to Arab Potash Company within 30 min by sinkhole-induced collapse of a saltpan dike in the Dead Sea region in 2000 (Arab Potash Company, 2000).

There are a variety of processes that can result in sinkhole development as well a wide range of scales that have to be considered in sinkhole research (e.g., Beck & Sinclair, 1986; Krawczyk & Dahm, 2011; Krawczyk, Polom, & Buness, 2015). It is this variability that requires different geophysical methods for optimal delineation of the structures and processes. During the formation stage, dissolution processes create cavities within a soluble host rock over periods of centuries to decades. Such regions of enhanced porosity and void space at depth form critical zones within rock units, which represent anomalies that can be identified by geophysical methods (Dahm, Heimann, and Bialowons, 2011; Dahm et al., 2010; Dobecki, 2010; Ezersky et al., 2013; Krawczyk, Polom, Trabs, et al., 2012; Kaufmann, Romanov, & Nielbock, 2011; Krawczyk, Polom, Dahm, 2012; Kühn et al., 2011).

The treat of surface subsidence originates from water access to soluble rock with the surface risk related to the size of the dissolution volume, its depth, and how it migrates to the ground surface.

SH–SH surveying has been used in sinkhole studies in the city of Hamburg, Germany (Krawczyk, Polom, Trabs, et al., 2012) (Fig. 23). The roof region of a salt diapir was studied that has long suffered dissolution and subsidence and most recently has become an area subject to microseismicity. Seismic data were acquired using similar system and setup as described for the quick-clay study in Sweden (Fig. 8). Small-scale structures in the sediments and salt were resolved to depths as shallow as 100 m with void migration and associated collapse structures at c. 60–80 m below surface. The fault structures interpreted on the seismic sections coincide well with the focal depth and mechanism established for the 2009 microseismic events. The largest of the recorded events was analyzed and located by Dahm et al. (2011) using data from monitoring stations some 100 m away. The source mechanism showed a strong normal faulting component and a moment magnitude of $Mw = 0.6$. The depth of the event was estimated at 80–120 m depth (Dahm et al., 2011).

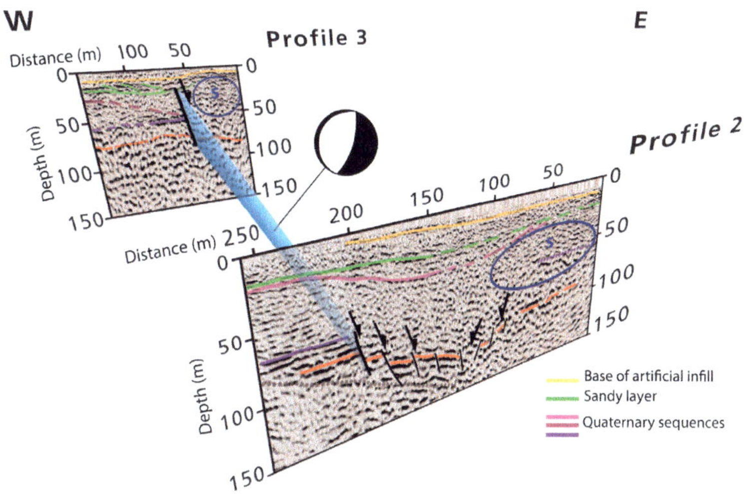

Figure 23 3D views showing interpreted SH–SH type reflection sections from a sinkhole area in the city of Hamburg. The location of the normal fault system imaged between 60 and 80 m depth corresponds well with the area where maximum microseismicity was felt in 2009. *Modified from Krawczyk, C.M., Polom, U., Trabs, S., & Dahm, T. (2012). Sinkholes in the city of Hamburg — new urban shear-wave reflection seismic system enables high-resolution imaging of subrosion structures.* Journal of Applied Geophysics, *78, 133–143.*

For most gradually subsiding sinkholes, migration of the solution void from the soluble host rock through the subsurface can take months to years before the first notable surface expressions. This migration process mainly affects the overburden and hydrological system. For instance, it is not uncommon to observe anomalies in the aquifer system or the occurrence of sporadic subsurface collapse earthquakes (Dahm et al., 2011; Lebert, Bernardie, & Mainsant, 2011; Malovichko, Dyagilev, Shulakov, & Butyrin, 2009; GISOS, 2009) during this migration phase.

Development of dissolution voids and associated subsidence features proceed through several stages based on hydrodynamics, stratigraphy, and overburden properties. Deformed overburden and abnormalities within the soluble rock interval interpreted on high-resolution seismic reflection sections have provided key insights into formation and development of dissolution voids and associated overburden subsidence processes (e.g., Miller, Villella, Xia, & Steeples, 2005; Miller, Xia, & Park, 2005). Subtle structures indicative of unique stages of overburden failure and rates can be identified within the collapse-altered volume on high fidelity and high signal-to-noise reflection sections (Fig. 24).

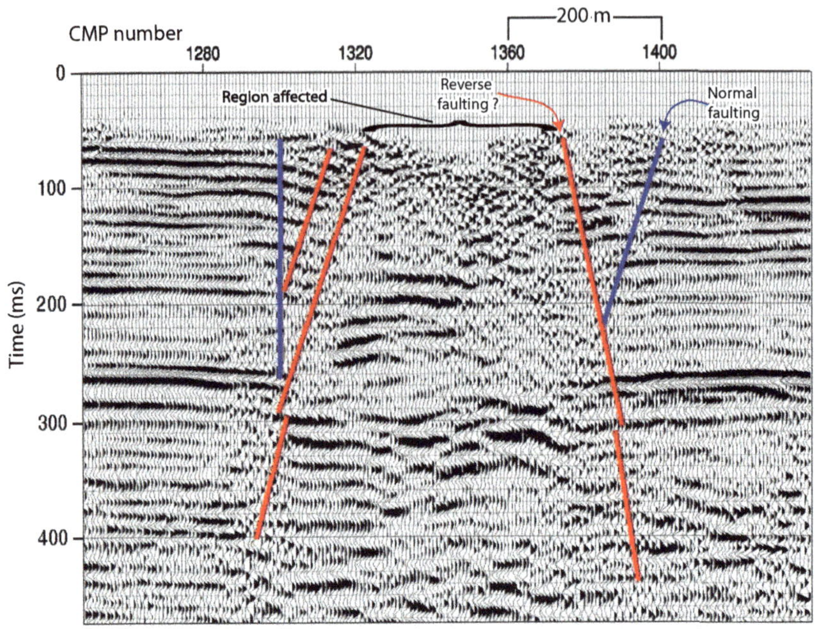

Figure 24 As a dissolution void migrates to the ground surface the upward progression of rock layers failures occur under compressional forces and therefore manifest themselves as reverse faults. Once the dissolution process has ended and forces move toward equilibrium the expansion of the throat and surface depression proceed under tensional forces and therefore normal faults are observed throughout the affected interval. This example is from central Kansas above the 125-m-thick Permian Hutchinson Salt and along a four-lane high-speed interstate highway (autobahn). Catastrophic failure could extremely be costly in lives and property.

Voids with a potential to affect the ground surface require a mined or dissolved volume height (common term in coal mining using room and pillar method or sublevel caving for metallic deposits) that is usually a little less than 10 times as deep as the layer thickness. This setting gives rise to the formation of crown holes at the surface. One or a combination of diverse drivers can be responsible for initiating void development and growth at a scale sufficient to exceed the strength of roof rocks, eventually migrating to the ground surface. In the majority of cases these mechanisms for void development include dissolution, natural and anthropogenic, of a bedded salt from oil field brine, mining, and erosion (Miller et al., 2010).

Soil cavities can form and slowly migrate upward and are normally highly dependent on hydrologic conditions (e.g., Waltham et al., 2005). Early recognition is important, but will only be possible and useful if the subsoil region is monitored at the cavity depth, with all applicable methods exploited and when possible incorporated with borehole techniques (Paillet, 2001) and numerical modeling (Kaufmann & Romanov, 2008; Kaufmann, Romanov, & Hiller, 2010). For instance, low-frequency seismic events were observed weeks to months before the Serville (France) collapse and sinkhole formation. These unique low frequency seismic events were associated with the breaking (fracturing and faulting) of marl layers at depth (GISOS, 2009). Bursts of long period seismic signals were also observed before the formation of the Bayou Corne sinkhole, Louisiana, United States, 2013 (e.g., Kent & Dunaway, 2013). Three-component extensometers and displacement transducers were deployed among other instruments in Thuringia/Germany to detect seismic energy associated with the collapse processes (Schmidt, Wunderlich, Peters, & Heinke, 2013).

The collapse process occurring during sinkhole formation can occur over days to hours. Monitoring thus requires high fidelity, continuous ground motion data, deformation, and possibly gravity measurements (e.g., Paine et al., 2009; Seidel & Serfling, 2010), in addition to geophysical and petrophysical characterization of overburden rocks and soil (Krawczyk, 2013a; Beilecke, Krawczyk, Ziesch, & Tanner, 2016). For instance, bursts of micro-earthquakes, long-period signals, and acoustic emissions, as well as accelerated void migration, have often been observed as precursors to final surface collapse events (e.g., GISOS, 2009; Kent & Dunaway, 2013; Lebert et al., 2011; Malovichko et al., 2009).

Modeling approaches aimed at defining geophysical imaging capabilities and limitations (Burschil, Beilecke, & Krawczyk, 2015; Burschil & Krawczyk, 2016), as well as process understanding and prediction (Kaufmann et al., 2010), have gained increasing importance, increasing confidence, and reducing the nonuniqueness of interpretations in the last decade. Modeling has also significantly improved our ability as scientists and engineers to communicate clearly to governmental decision-makers or insurance companies (Krawczyk, 2013a, 2013b). Physical models have provided insights into the process and progress (Ge & Jackson, 1998), and numerical models (Chen & McMechan, 1993) provide a greater understanding as to the imaging limitations and expectations with real-world problems.

High-resolution seismic-reflection data have a proven track record of accurately imaging subsidence features both before and after a subsurface void migrates to the surface to form a sinkhole. The uniqueness of the challenges in the near surface relative to deeper imaging objectives cannot be taken for granted and requires a tuned acquisition and processing approach with care taken during the selection of even the most basic parameters. Changes in the velocity field within the dissolution and subsidence volume across subspread distances can require velocity analysis be undertaken as a function of offset ranges and time (Miller et al., 1995). Many times this level of accuracy requires the splitting of a single CMP (common mid-point) gather into offset subsets.

Dissolution from anthropogenic or natural fluid sources results in a wide range of overburden subsidence structures with seismic representations that provide clues but no definitive interpretations as to the complete development history. An underlying difference between anthropogenic (Fig. 25) and natural (Fig. 26) dissolution processes and resulting structures that affects seismic interpretations of current and past collapse development is the one-dimensional hydrodynamic system driving the anthropogenic process compared to the three-dimensional process for natural void and subsidence features.

Several key aspects critical to accurate seismic imaging of subsidence structures must be considered or interpretations can significantly diverge

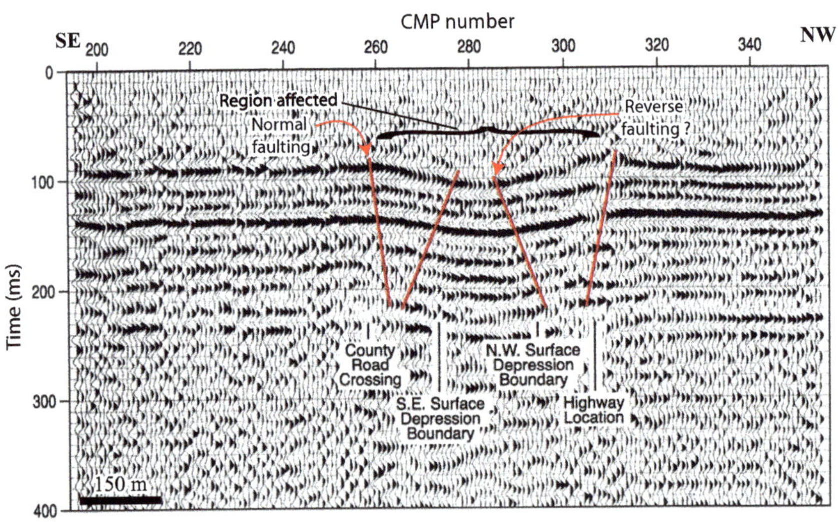

Figure 25 CMP P-wave reflection stacked section across a sinkhole that formed gradually and is centered on an oil field brine disposal well. A breach in the disposal well casing allowed unsaturated brine to gain access to the Permian salt beds around 400 m below ground surface and more than 1 km above the intended disposal interval. The wellbore acted as both the source and escape point for the disposal fluids. This single uniform synform imaged on this seismic section from central Kansas, United States, is a classic representation of an anthropogenic dissolution feature.

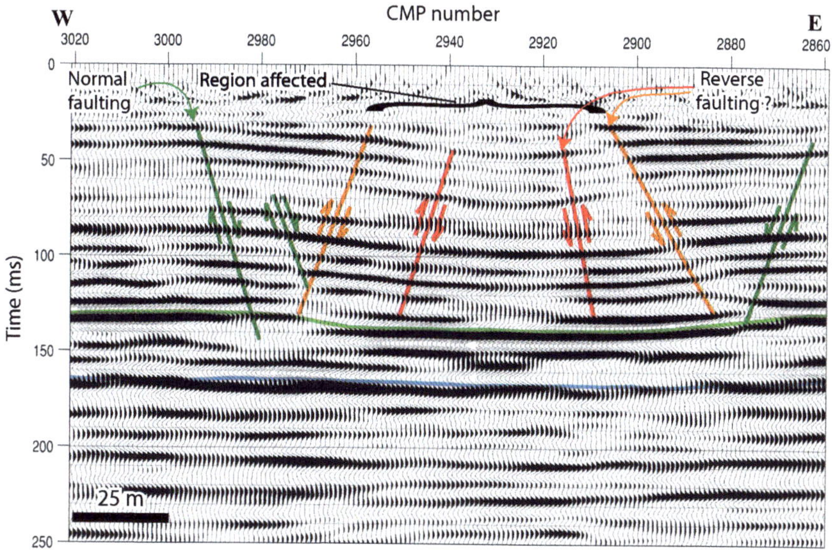

Figure 26 CMP stacked section across the natural dissolution edge of the Permian Hutchinson Sal in central Kansas, United States. The distinctive erratic and undulating nature of the subsidence structures is very consistent with the multiple inlets and outlets for fluids that can naturally gain access to the salt along this receding edge of the 125-m-thick salt interval.

from the reality in the subsurface. Interpretations of reflections below a subsidence feature suffer from subsolution zone static and out-of-plane energy generally masking reflections in the first few hundred milliseconds below the soluble rock interval. Two-dimensional seismic surveys must carefully consider line locations when imaging these small (relative to wavelength) geometrically irregular features to provide accurate interpretations. Classical horizontal resolution-limits (as defined by the Fresnel zone concept) appear to not accurately represent true potential based on seismic images of steep-sided collapse structures. Seismic investigations using 3D techniques are critical to continued development of accurate dissolution; creep; and/or subsidence scenarios and processes based on empirical, numerical, and physical models.

3.5 Earthquakes

Strategy for seismic risk mitigation relies on the mapping of seismic hazard, estimation of seismic site response, and knowledge-driven urban planning and construction design. In high seismicity countries, site surveying is needed to estimate the effect of morphological and stratigraphic settings on local ground motion (Albarello, Socco, Picozzi, & Foti, 2015). Central authorities often provide seismic hazard estimates at a national level. Near-surface seismic characterization can provide different and key information in the context of site-specific response. In particular, S-wave velocity (V_S) models and quality factor can be estimated and used as input for seismic site response evaluation. This is true both in simplified and complex seismic conditions where morphology and subsurface discontinuities can aggravate local damages due to amplification and dynamically triggered instabilities (liquefaction, induced landslides) (e.g., Tertulliani, Leschiutta, Bordoni, & Milana, 2012). Essential information can range from simple identification of soil classes, often defined as a function of the parameter V_{S30} (the time average S-wave velocity down to 30 m depth) as requested by many regulatory codes (e.g., Eurocode 8, 2003) to 3D-V_S models and laboratory characterization.

In many cases the local seismic response is evaluated through 1D modeling (e.g., Kramer, 1996) requiring a 1D-V_S profile and strain-dependent dynamic parameters that describe soil response (stiffness and damping ratio decaying curves, bulk density). The 1D-V_S profile is usually estimated through borehole seismic measurements (down-hole or cross-hole tests) or surface-wave analysis. The depth of interest can range from few tens of meters to hundreds of meters. To sample this large depth range active and passive surface-wave data are often acquired and merged into a single profile.

A lively debate has been ongoing within the scientific community to evaluate the adequacy of surface-wave analysis results for incorporation into local seismic response modeling and microzoning studies.

Several studies have been performed to estimate the propagation of uncertainties of the V_S model on the seismic response parameters (Boaga, Vignoli, & Cassiani, 2011; Comina & Foti, 2015; Foti, Comina, Boiero, & Socco, 2009; Jakka, Roy, & Wason, 2014; Roy, SankarJakka, & Wason, 2013; Socco, Foti, Comina, & Boiero, 2012) and on the robustness of estimation of time average velocity (Brown, Diehl, & Nigbor, 2000; Comina, Foti, Boiero, & Socco, 2011). Moreover several blind tests have been conducted on both passive and active data (Cornou et al., 2009; Cox, Wood, & Teague, 2014; Tran et al., 2011; Xia et al., 2000). Among others, the InterPacific project, recently concluded, deserves some attention due to the worldwide participation of research teams and surveying companies (Garofalo et al., 2016a, 2016b).

For surface-wave analysis (Garofalo et al., 2016a), the results submitted by 14 international teams showed very little variability in the experimental dispersion curves estimated with different processing approaches. This finding confirms that the many different processing algorithms currently in use and accepted by the scientific community all provide robust and stable results. Most of the variability in the estimate of the S-wave velocity profile is therefore associated to parameterization of the reference layered model and solution nonuniqueness. The extensive comparison between in-hole seismic methods and surface-wave analysis at three different sites with different geological conditions (Garofalo et al., 2016b) has shown that although the former provide a higher spatial resolution with depth, the variability in the estimate of the S-wave velocity as a function of depth is comparable between the two classes of methods.

When a local 1D approach does not adequately provide a reliable estimate of the expected ground motion, 3D-V_S models are required at basin scale. In the building of 3D-V_S models all the available knowledge coming from geological surveys, borehole measurements, and surface-based seismic measurements are integrated to ensure that the required spatial resolution and investigation depth are met. Luke et al. (2008) presented a velocity model of the entire Las Vegas Valley that was obtained by integrating geological mapping, borehole information, and a large set of surface wave local surveys. Various available measurements and interpolation of local 1D-V_S profiles helped to obtain a regional 3D model of the study area. An alternative to this approach is to invert simultaneously a set of local dispersion curves for local 1D models using a laterally constrained inversion scheme (Auken & Christiansen, 2004; Socco et al., 2009) in which the model parameters of the different 1D profiles are linked with each other. This is accomplished by using a set of spatial constraints scaled by the mutual distance among models and whose strength depends on the expected lateral variability at the site. *A priori* information from borehole measurements and geological mapping can also be included in the inversion in terms of constraints whose robustness depends on the reliability of the information itself.

Piatti, Foti, Socco, and Boiero (2013) have applied this approach for seismic characterization of Tarcento basin (an area in northeast of Italy with high seismicity). The city of Tarcento was affected in 1976 by two earthquakes (known as Friuli earthquakes, from the name of the Italian region) that collectively resulted in 965 fatalities across this region. Damage in the Tarcento area as a result of these earthquakes was attributed to local site amplification effects related to seismic velocity heterogeneities. The basin is

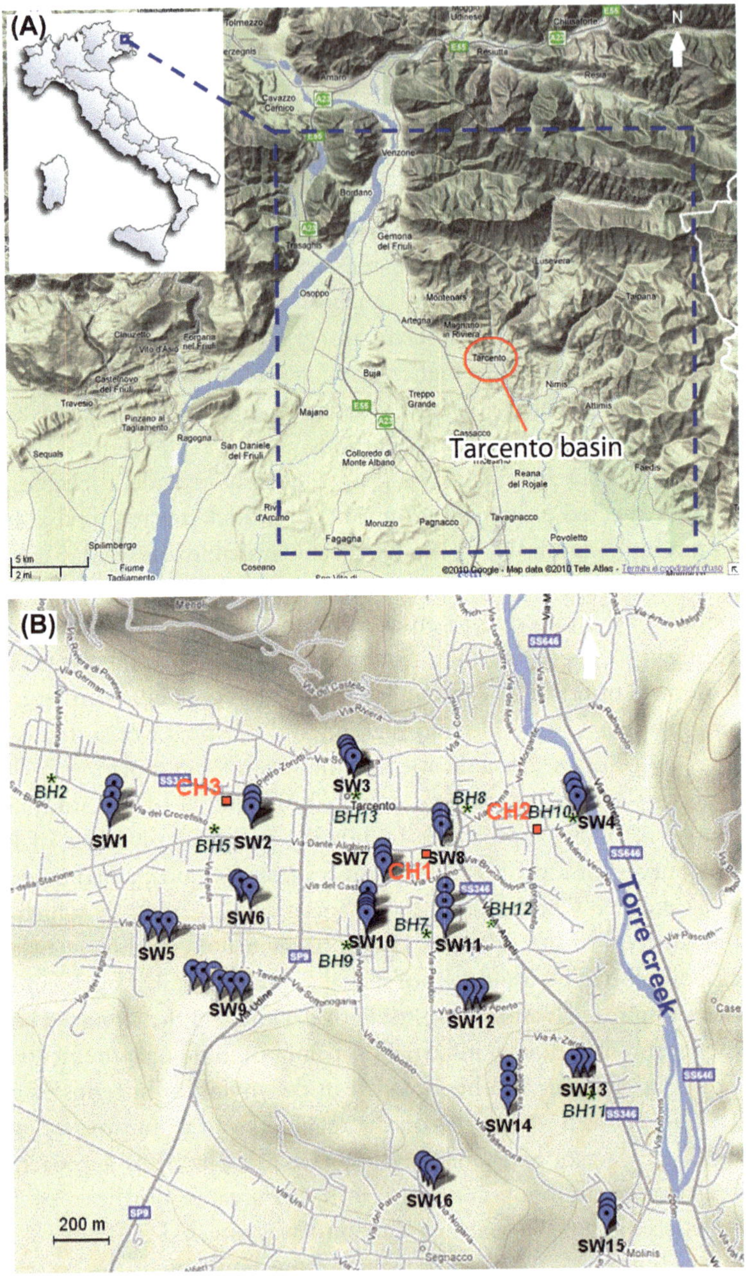

Figure 27 (A) Locations of Tarcento basin and (B) seismic stations used for surface-wave data inversion as well as vintage boreholes. *Modified from Piatti, C., Foti, S., Socco, L. V., & Boiero, D. (2013). Building 3D shear velocity models using surface wave testing: the Tarcento Basin case history:* Bulletin of the Seismological Society of America, 103*(2A), 1038–1047, http://dx.doi.org/10.1785/0120120089.*

bounded by mountains along its northern, eastern, and southern boundaries (Fig. 27A), and the seismic bedrock is a flysch formation consisting of alternating layers of marls, sandstones, and conglomerate overlying a deep limestone formation. The thickness of the alluvial deposits (gravels and sands, with rare layers of silts and clays) overlying the bedrock varies significantly over the basin from almost zero to more than 100 meters.

Piatti et al. (2013) used vintage data including nine geomechanical sounding and three cross-hole tests (Cauzzi et al., 2011), and a new active and passive surface-wave measurements performed at 16 sites across the whole Tarcento basin (Fig. 27B). The surface-wave data were principally intended to estimate V_S of the alluvial deposits and determine the bedrock depth and impedance contrast at sites where the bedrock is within the depth of investigation range for surface wave methods (on average less than about 50 m). Borehole information and outcrop studies were used to create *a priori* information about the bedrock depth to be included in the spatially constrained inversion (SCI) of surface-wave data. These data were used to complement the information where the bedrock depth was beyond investigation depth. A smooth final V_S model over the whole basin was constructed and used for 3D modeling of the seismic response of the basin (Fig. 28).

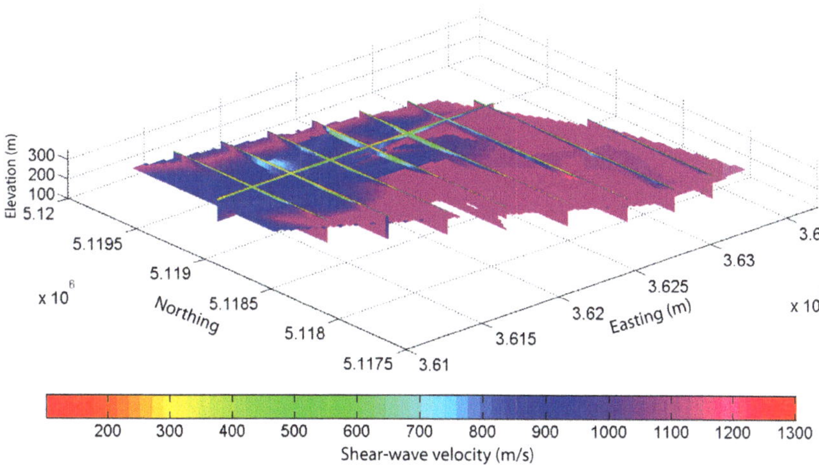

Figure 28 Shear-wave velocity model obtained from surface-wave data in the Tarcento basin in Italy. *Modified from Piatti, C., Foti, S., Socco, L. V., & Boiero, D. (2013). Building 3D shear velocity models using surface wave testing: the Tarcento Basin case history:* Bulletin of the Seismological Society of America, 103*(2A), 1038–1047, http://dx.doi.org/10.1785/0120120089.*

Another important target in seismic hazard studies is the location of active faults that affect the shallow layers up to depths very near the ground surface but do not outcrop or can be inferred from surface expressions (blind faults). Mapping of faults can help modeling the seismicity of an area and can also be useful in hydrogeological studies since faults can act as barriers to groundwater flow. Faults can produce solid matrix structure modifications along disaggregation bands in shallow, soft sediments where intrusion of liquefied sand along weak zones produce step-like undulation of layer boundaries and the creation of barriers to fluid (Ghose, Carvalho, & Loureiro, 2013). Modifications induced by the deformation caused by the fault can then be recognized as variations in the seismic properties of the soil. For these reasons, seismic methods, in particular high-resolution seismic reflection and surface wave analysis together with direct surveys (trenching and coring) can be very effective in delineating and characterizing faults in shallow layers.

In spite of the fact that high-resolution P-wave reflection seismic method has been successful for delineating shallow faults, when fault displacement is small or when the fault plane is subvertical, reflection seismic methods commonly do not produce the desired results. In some of these cases, S-wave velocity, which is more sensitive to soil strengths, can provide some useful information. Ghose et al. (2013) identified shallow faults in soft sediment at Vila Franca Xira (Portugal) using velocity analysis and modeling of prestack SH-type high-resolution data. Hayashi (2014) provided a comprehensive seismic model of East San Francisco Bay by integrating a wide set of active and passive surface wave surveying.

Malehmir et al. (2016) recently used a combination of RMT, ERT, refraction seismic, GPR, gravity, and magnetic methods to identify and characterize a surface scarp that was associated with a multiphase postglacially reactivated fault in central Sweden (Fig. 29). The various geophysical methods complemented each other and allowed for a well-constrained interpretation of the lidar imaged scarp. The GPR survey showed a reflection calculated to be 4–5 m deep and not detected by any of the other methods, while the bedrock surface (10–20 m depth) was resolved by several other methods. This fault was principally inferred from a low-velocity zone detected and imaged with first break traveltime tomography methods (Tryggvason, Rögnvaldsson, & Flóvenz, 2002) and a high-conductivity interval interpreted from ERT and RMT data. RMT was superior in resolving the conductive zone relative to ERT, and ERT provided a high-quality representation of the top of bedrock.

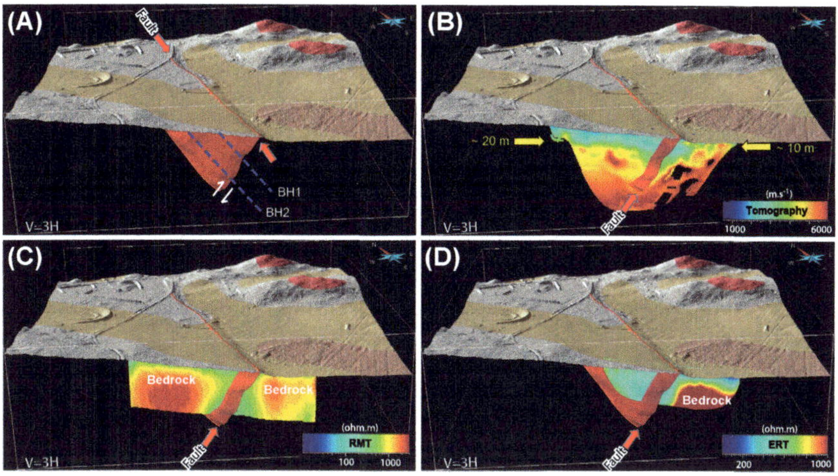

Figure 29 3D visualization of the geophysical results showing (A) the surface expression of a reactivated postglacial fault on the lidar data with a hypothetical shape of the fault plane (assumed to be reversed) and partly constrained from the measured magnetic data, (B) a low seismic velocity zone associated with the fault indicating fractured rocks in the fault zone, (C) radio magnetotelluric (RMT)-, and (D) electrical resistivity tomography (ERT)-conductivity models suggesting a major conductivity zone in the bedrock indicating fractured and water-bearing rocks related to the fault. *Modified from Malehmir, A., Andersson, M., Mehta, S., Brodic, B., Munier, R., Place, J., ..., Lund, B. (2016). Post-glacial reactivation of the Bollnäs fault, central Sweden — a multi-disciplinary geophysical investigation.* Solid Earth, *7, 509—527.*

Controlled-source (1—250 kHz) RMT surveys were also used in the Volvi basin of Thessaloniki, Greece, to delineate an active fault and appraise the potential for damage due to seismic amplification caused by thick overburden materials (Bastani, Savvaidis, Pedersen, & Kalscheuer, 2011) (Fig. 30). A sudden drop in a resistivity high representing some form of truncation is evident on most profiles and was interpreted to represent the location of the fault. Additional faults were also inferred to be present based on similar characteristic evident in these data. Various depth slices, produced by 3D gridding of the 2D profiles allowed confident mapping of the location/trace of the fault zone that separates a region of high resistivity (bedrock contact) from low ones (overburden sediments) (Fig. 31).

An alternative approach to infer fault locations is through the analysis of surface-wave energy that has back reflected from the subvertical discontinuity represented by the fault. Hyslop and Stewart (2015) showed on synthetic and field seismic data acquired across the Hockley Fault near Houston, Texas, that faults can be identified by building surface-wave

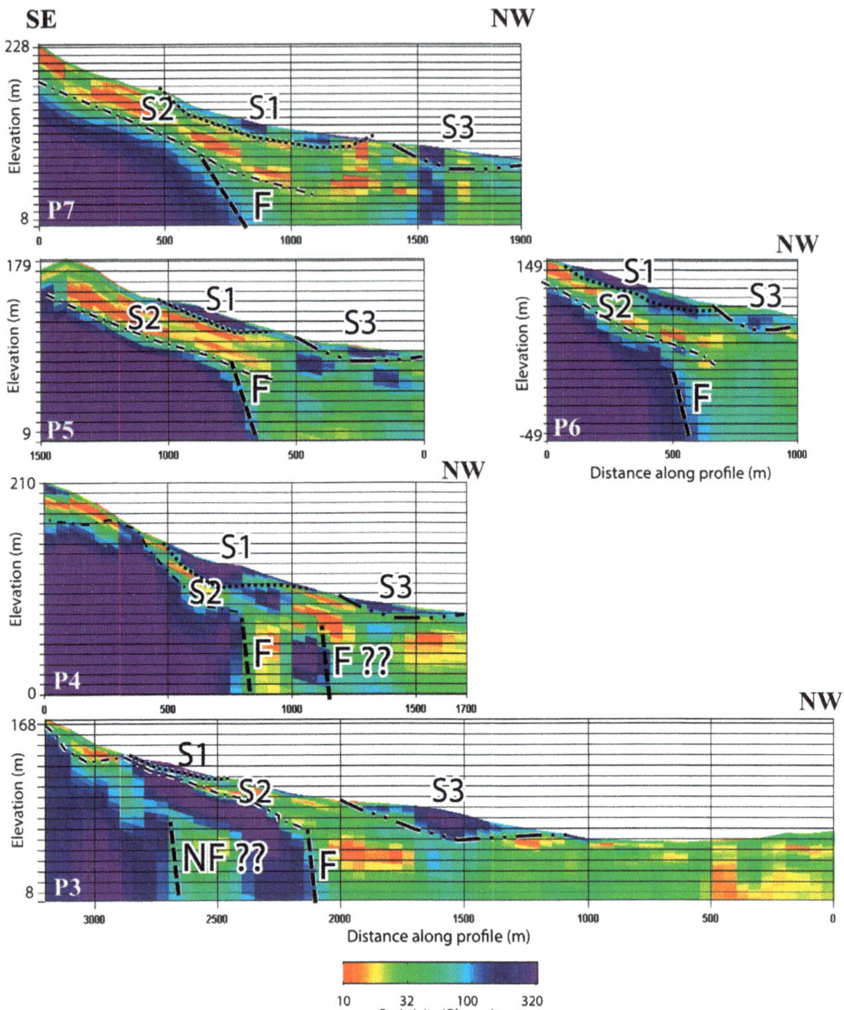

Figure 30 2D models of resistivity from controlled-source radio magnetotelluric (CSRMT) data acquired in the Volvi basin in Greece. A fault was inferred to be located on the north-western boundary of the high-resistivity zone (labeled as F). Other features such as S1—S3 were interpreted to be associated with the overburden materials. *Modified from Bastani, M., Savvaidis, A., Pedersen, L.B., & Kalscheuer, T. (2011). CSRMT measurements in the frequency range of 1—250 kHz to map a normal fault in the Volvi basin, Greece.* Journal of Applied Geophysics, 75, *180—195.*

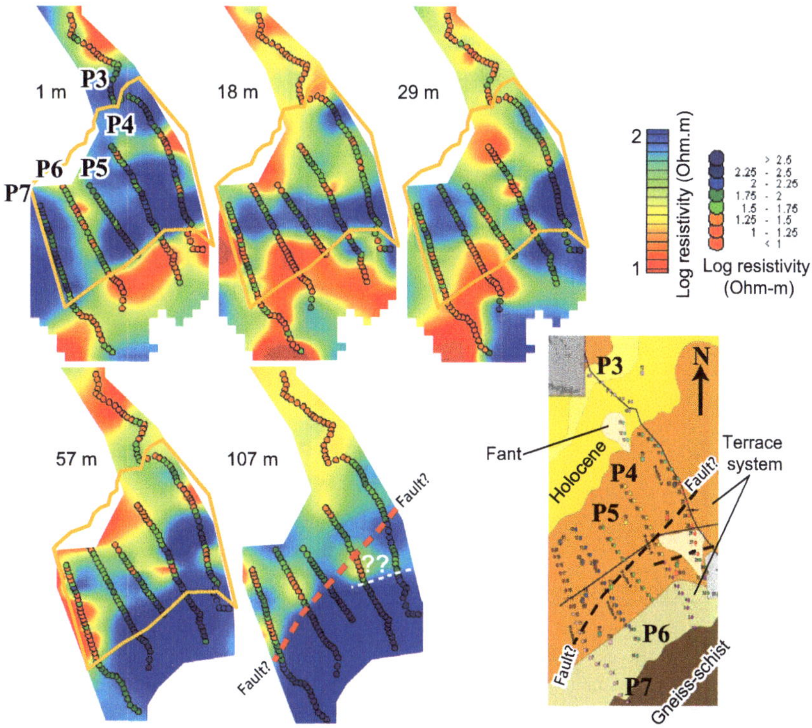

Figure 31 Resistivity depth sections (plan view) produced by gridding and interpolating from the 2D results showing possible location of the fault. *Modified from Bastani, M., Savvaidis, A., Pedersen, L.B., & Kalscheuer, T. (2011). CSRMT measurements in the frequency range of 1–250 kHz to map a normal fault in the Volvi basin, Greece.* Journal of Applied Geophysics, *75, 180–195.*

reflectivity sections. Their method is similar to the computation of the austospectrum of the seismic traces proposed by Zerwer, Polak, and Santamarina (2005) for the detection of cracks in concrete beams and applied to fault detection by Bergamo and Socco (2014). Bergamo and Socco (2014) also extended the method based on attenuation analysis of Rayleigh waves developed by Nasseri-Moghaddam et al. (2005) for the estimation of the location and depth of underground cavities. The method was applied to multifold seismic data acquired over the New Zealand Alpine fault (see Carpentier et al., 2010, 2012 for details about the data and the site).

The method computes the decay exponent of Rayleigh waves, which represents attenuation or amplification along the line. Energy back reflected by a vertical discontinuity interferes with the forward propagating wavefield

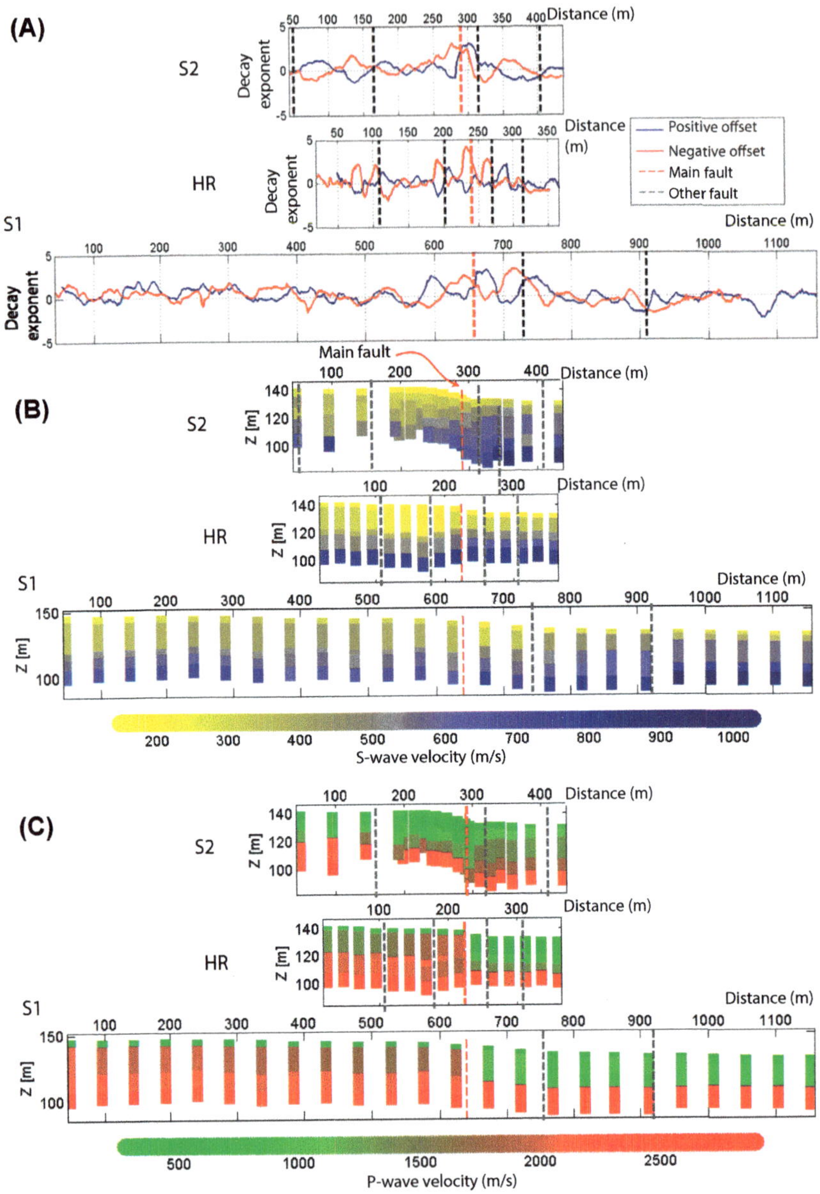

Figure 32 Fault delineation over three seismic lines (S1, S2, and HR) crossing the Alpine fault in New Zealand. (A) Surface-wave decay exponent for positive and negative offsets, (B) V_S, and (C) V_P models along the lines from joint inversion of surface-wave dispersion curves and P-wave first breaks.

and can produce a local increase of energy (amplification). Beyond the discontinuity (opposite side of the fault the source is on), the energy is lower than predicted due to losses related to the back-reflected waves. With a seismic line laid across the fault and source points located on both sides of the fault (positive and negative offsets), the intersection of the energy decay exponent plots calculated for each source along the line corresponds to the position of the fault (Bergamo & Socco, 2014).

From the same data set Konstantaki, Carpentier, Garofalo, Bergamo, and Socco (2013) extracted surface wave dispersion curves and inverted them with a laterally constrained inversion algorithm to obtain V_S sections along five seismic lines that crossed an Alpine fault. Fault locations from cross sections obtained using the method proposed by Bergamo and Socco (2014) compared to the results of the joint inversion of surface-wave dispersion curves and P-wave traveltimes (Boiero & Socco, 2014) for three seismic lines are very similar (Fig. 32). The interpreted position of the main fault (red dashed line) matches the results from seismic reflection data presented by Carpentier et al. (2010); minor faults are also identified by the study (Fig. 32). Some lateral variations in velocity are visible in the V_P and V_S models corresponding to the fault locations. The lateral resolution of the fault detection method is strictly related to the geophone spacing (2 m in line S1 and S2 and 1 m in line HR), while for the joint inversion of surface-waves and P-wave traveltimes the lateral resolution depends on the length of the used spatial window which is 96 and 48 m, respectively.

3.6 Volcanoes

Volcanoes can directly and indirectly pose a variety of hazards (see Grgich, 2013) including (1) lava flows overrunning roads and towns (e.g., historical Pompeii in Italy), (2) ash falls blanketing the landscape (e.g., in Chile 2015), (3) pyroclastic flows incinerating towns and fields, (4) landslides and lahars burying land surface (e.g., in Guatemala and Indonesia), (5) earthquakes topple structures and rupturing dams (Roman & Cashman, 2006), (6) tsunamis washing away coastal towns, (7) individual gases suffocating nearby people and animals, and (8) ash clouds perturbing plane traffic as mentioned earlier. Volcanic hazards are typically evaluated and characterized through the study of magma movement within the crust, changes in hydrothermal activities, and edifice stability. Passive seismic monitoring was the earliest geophysical method used to anticipate volcanic eruption relating increased seismicity to magma movement.

Passive seismic monitoring is the principle and still the only monitoring technique on many active volcanoes but methods and sensors have improved and diversified. Arrays of short-period seismometers are used to detect and locate earthquakes in the volcano edifices (e.g., Rymer, Ferrucc, & Locke, 1998). Broadband seismic arrays (permanent or on rapid-response campaigns) have recently become popular for real-time mapping of fluid movement in thc crust (Chouet et al., 2003). Infrasound microphone systems are also used to record the volcano-acoustic signals, for example, tremors from resonating dikes and conduits (e.g., Curilem, 204). The fundamental frequency of a tremor is related to the volume of the resonating body and changes with the properties of the magma and its inherent gas load and state. Acoustic flow monitoring measure ground vibration caused by lahars or debris flows (USGS, 2016a). They are mostly used as early warning systems for evacuations.

Additional techniques such as geochemistry, temperature, and deformation monitoring have become an option at many locations. Technologies available to monitor deformation due to volcanic activities are similar to those used for landslide monitoring and include approaches such as satellite remote sensing including InSAR, leveling, and electronic distance meter (EDM) surveys, direct measurement and telemetered GPS networks, lidar investigation, electronic tiltmeters, and strainmeters.

Near-surface geophysical surveys have utility for imaging the overall and sometimes internal architecture of volcanoes. Active seismic surveys have been used to reconstruct detailed shallow crustal structures beneath active volcanoes and for delineating their feed systems. An early 2D seismic tomography study of Mt. Vesuvius (Zollo et al., 1998) encouraged many similar surveys. ERT and self-potential surveys can image hydrothermal fluid circulations in a volcanic edifice and thereby map the geometry of the hydrothermal system, as well as reveal potential failure planes (Brothelande et al., 2014). AEM surveys can rapidly cover entire volcanic edifices and image their architecture. The data quality is usually reduced by abrupt topography but advanced processing algorithms can overcome this (Dumont et al., 2015).

3.7 Cryospheric Hazards

Cryospheric hazards occur in high mountainous as well as high latitude areas, such as the Arctic and Patagonia, all over the globe. Their impact on society, however, depends on the density of human presence in those regions. Ice loss, glacier lake outburst floods, avalanches, rock-ice landslides, lahars can lead to dramatic or even catastrophic impacts on regional populations (Haeberli & Whiteman, 2014). Glacier- and permafrost-related mass movements, permafrost thaw settlements, and frost heave are additional hazards within the cryospheric zone. These hazards require more and more attention due to their impact of climate change in these sensitive environments. Kääb (2005) and Kääb et al. (2005) provide comprehensive overviews on feasible remote sensing technologies for monitoring glacier- and permafrost-related hazards.

3.7.1 Rock Glaciers

Rock glaciers are the result of mountain permafrost creep and as such, are common features in high-alpine periglacial areas. Helicopter-borne GPR surveys are feasible (Blindow, Salat, Gundelach, Buschmann, & Kahnt, 2011; Rutishauser, Maurer, & Bauder, 2016) and usefulness for the monitoring landforms of that type and their potential hazards. To demonstrate the potential of that method, we present a case study recently performed on mountain permafrost region, Furggwanghorn of Turtmann valley, in Switzerland (Fig. 33A). All the technical and interpretational details of this

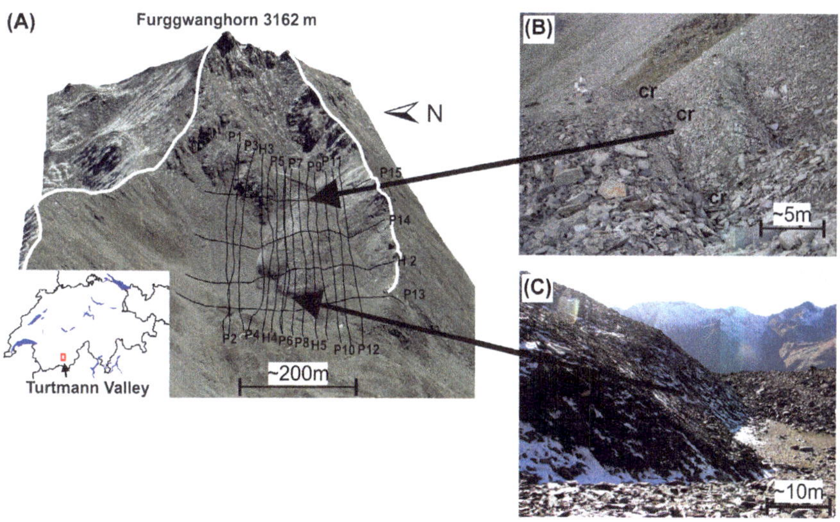

Figure 33 (A) Furggwanghorn rock glacier showing the helicopter-GPR profiles. (B) Crevices (cr) in the root zone and boundary surface, and (C) steep frontal slope in the study area justifying the use of helicopter GPR survey. *Modified from Merz, K., Green, A.G., Buchli, T., Springman S.M., & Maurer, H. (2015). A new 3D thin-skinned rock glacier model based on helicopter GPR results from the Swiss Alps.* Geophysical Research Letters. *http://dx.doi.org/10.1002/2015GL063951.*

study are documented in Merz, Green, Buchli, Springman, and Maurer (2015), Merz, Maurer, et al. (2015), and Merz et al. (2016).

The Furggwanghorn rock glacier lies at altitudes between 2750 and 2900 m above sea level. The area is part of the crystalline alpine nappes and the geology consists mainly of schist and gneisses. Weathering of these rocks has produced materials ranging from silt to angular shaped blocks up to 2 m in diameter (Fig. 33B). This rock glacier is distinguished from others by large movement rates of up to several meters per year. As a result of these large movements, deep surface depressions have formed near the root zone of the rock glacier (Fig. 33B).

Geophysical characterization of the Furggwanghorn rock glacier was complicated by several factors: (1) logistically challenging and expensive to transport equipment and personnel to the site, (2) the surface of the rock glacier is extremely rugged, making it difficult to move sensors along profiles on the ground, and (3) some areas were inaccessible due to safety concerns (Fig. 33C). After considering all these factors it was judged to be worthwhile to attempt surveying the Furggwanghorn rock glacier with helicopter-borne GPR. Standard GSSI data acquisition system was equipped with a pair of Radarteam Subecho-40 antennas (~70 MHz dominant frequency). This

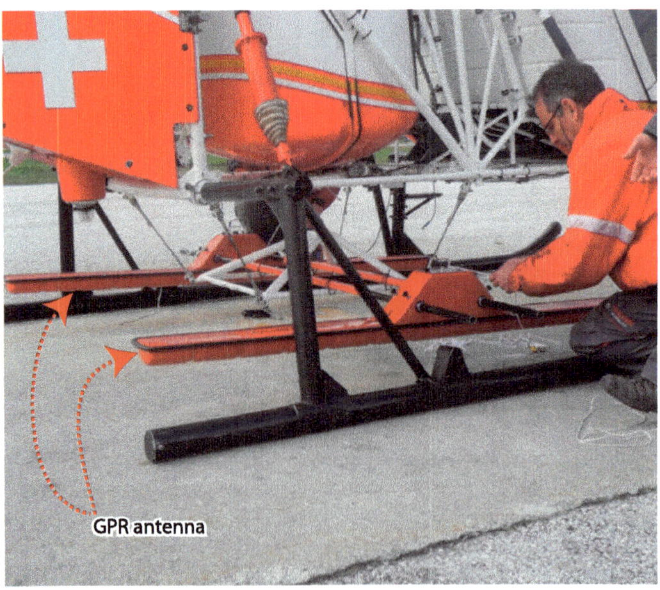

Figure 34 Ground-penetrating radar antennas mounted underneath the helicopter.

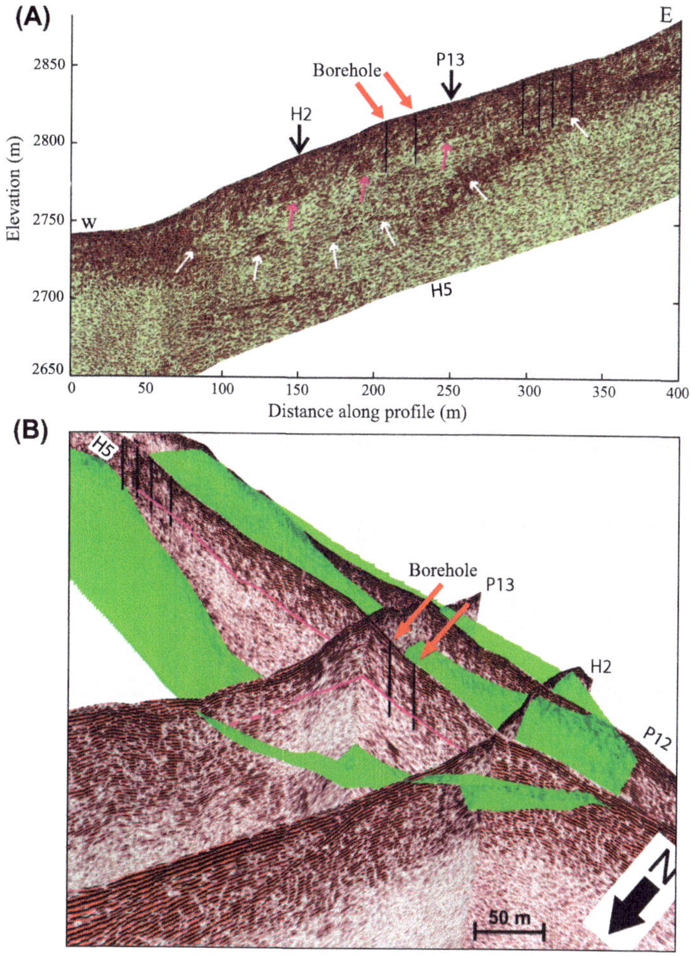

Figure 35 (A) Helicopter ground-penetrating radar (GPR) profile H5 (for location see Fig. 33A). Topography is taken into account in the display. *Arrows* mark crossing points of profiles H3 and P13. Locations of the boreholes are shown with *black lines*. The depth of the interpreted basement is indicated with *white arrows*, and transition of ice-rich to ice-poor rock glacier material is marked with *magenta arrows*. (B) Fence diagram of helicopter GPR profiles H2, H5, P12, and P13. The interpreted bedrock is shown as green surface and the boundary between ice-rich and ice-poor permafrost is marked in magenta. *Based on Merz, K., Maurer, H., Rabenstein, L., Buchli, T., Springman, S.M., and Zweifel, M. (2016). Multidisciplinary geophysical investigations over an alpine rock glacier.* Geophysics, 81, *WA147–WA157.*

provided a spatial resolution of about 2 m. The antennas were mounted beneath the helicopter (Fig. 34), and the control console was placed inside of the helicopter. Accurate positioning of the surveying points within the centimeter range was obtained using a DGPS system.

A grid of 20 profiles was acquired in about 1 h (Fig. 33A). A comparable ground-based survey would have required several days, and some areas would have been impossible to cover with ground-based GPR. A relatively simple processing sequence was applied, including (1) removal of harmonic noise (originating primarily from reflections from the helicopter) using a singular value decomposition filter, (2) frequency filtering, (3) altitude corrections, (4) data binning, (5) fx-deconvolution, (6) time migration, and (7) time-to-depth conversion (see Merz, Green, et al., 2015 for details) (e.g., Fig. 35A and H5 in Fig. 33A).

The interface between the bedrock and the rock glacier material is clearly visible and possesses a maximum depth of about 50 m (white arrows) (Fig. 35). Additionally, a characteristic change in the reflection pattern is evident at about 20 m below surface (magenta arrows in Fig. 35A). This was interpreted as the transition from a shallow ice-rich layer to an underlying ice-poor zone. At this interface, almost all of the deformation occurs in a focused shear zone (Buchli, Merz, Zhou, Kinzelbach, & Springman, 2013). Therefore, knowledge of the geometry and depth of the shear zone is key for assessing the stability of a rock glacier. Boreholes drilled into bedrock confirmed bedrock depth and the presence of an ice-rich/ice-poor interface at about 20 m depth (Buchli et al., 2013; Merz et al., 2016).

The bedrock interface and ice-rich/ice-poor transition zone were observed on most profiles, which allowed their 3D shape to be determined (green surface, Fig. 35B). Knowledge of the 3D bedrock topography allowed estimation of the volume of the Furggwanghorn rock glacier at 910,000 m^3. These results form an important prerequisite and a starting point for a geotechnical characterization of the Furggwanghorn rock glacier, which is currently underway.

In addition to the permafrost study discussed in this paper, promising initial attempts with helicopter-borne GPR systems have been undertaken in landslide areas. We expect this technique has considerable potential for other applications as well, especially where accessibility to the surface is difficult or challenging.

3.7.2 Avalanches

Eckerstorfer, Bühler, Frauenfelder, and Malnes (2016) recently published a review of remote sensing applications for snow avalanche characterization, mapping, and monitoring while highlighting recent advances, as well as reasonable potentials and limitations of existing techniques. There are a variety of geophysical methods that can address potential issues associated with snow avalanche hazards. It is well understood that the key parameters to capture are snow pack thickness, internal layering within the snow pack, and the general slope of the area of concern. Unfortunately, remote methods for reliable assessment of the snow cover stability are still missing from our geophysical repertoire.

4. ROAD AHEAD

More attention than ever is being given to conventional near-surface geophysical methods and applications for geohazard studies. Technological and methodological improvements allow unprecedented opportunities to acquire higher quality data in a more time- and cost-effective manner than ever. More and more integrated geophysical methods combined with geotechnical or geomechanical measurements are implemented to reduce ambiguity from one method and to provide constrained interpretation of an individual method (e.g., COST, 1995; Kottmeier et al., 2016; Krawczyk & Dahm, 2016; Salas-Romero et al., 2016; USGS, 2016b). Remotely operating vehicle (ROV) and autonomous underwater vehicles (AUV) systems equipped with many geophysical sensors and surveying systems are allowing access to difficult and affected terrains while surface-based and borehole methods gain from access to the most advanced technologies. Here we provide potential improvements on some of the methods while keenly aware that much more can be said and addressed than we have here and in greater details.

4.1 Multicomponent 2D and 3D Active and Passive Seismic Surveys

While most examples provided here use 2D approaches to study a particular geohazard, in most cases the problem is complex and 3D making an improved and accurate understanding only possible using 3D approaches (Heincke et al., 2006; Malehmir & Bellefleur, 2009). As illustrated here through various case studies, use of shear-wave energy for imaging and subsurface characterization is becoming quite popular for many applications (Garotta, 2000; Ghose, 2002). However, the best approach is likely to be a combination of all wave-modes as we strive for the most accurate imaging and characterization possible. SH—SH surveys can yield high-resolution image when there is sufficient structural continuity and a well-suited velocity structure in the subsurface (Krawczyk, Polom, & Beilecke, 2013). However, there are many settings, such as when fresh volcanic or slump materials are associated with the target being studied, that partnering with other approaches might be the best path forward. In this case, P-wave refraction tomography or MASW can prove useful.

Seismic technologies involving broadband accelerometers (e.g., microelectromechanical sensors) have the benefit of acquiring a full spectrum of data and doing so, allow optimal use of both low-frequency imaging methods (such as full-waveform tomography and MASW) and high-resolution reflection imaging (e.g., Hons, Stewart, Lawton, Bertram, & Hauer, 2008). When these sensors are multicomponent (e.g., 3C), recorded data can be used collectively and processed for operations such as full waveform inversion or separated into each mode (e.g., P—P, SH—SH, and SV—SV as well as mode converted P—S and S—P). P—P and SH—SH are currently the most popular because of their simplicity for imaging and unique propagation characteristics. Brodic et al. (2015), Malehmir, Zhang, et al. (2015), and Malehmir, Wang, et al. (2015) report examples of such a system to address issues with urban applications and how cost-effectively they can be used when combined with wireless recorders. Acquisition systems that accommodate both passive and active data recording will see escalating use as data integration continues to advance and improve.

On the technology side, fiber optics and other novel instrumentation designs will continue to be used for structural monitoring and change in internal stress and strain such as for landslides or rockfall applications (Harp et al., 1995; Moore, Gischig, Button, & Loew, 2010) as well as for imaging purposes (e.g., Bostick, 2000; Hornby, Bostick, Williams, Lewis, & Garossino, 2005). They are currently being used and tested in various permanent monitoring applications particularly for CO_2 sequestration and reservoir monitoring (e.g., Daley et al., 2013; Jousset, Reinsch, Henninges, Blanck, & Ryberg, 2016).

In terms of technical developments, joint inversion and modeling and noise interferometry approaches are going to become popular for geohazard applications because of their potential in noisy and populated places. Joint inversions will allow models that honor various data sensitivities and satisfy properties of various data sets. In so doing, multidata type inversion routines will likely be more reliable than individual models (depending how it is formulated). Place and Malehmir (2016) have, for example, recently used virtual first break interferometry to improve the accuracy of first break picks for both tomographic applications and refraction static solutions for reflection imaging. Their approach, for example, allowed a better delineation of a multiphase deformation zone that could put infrastructure at risk during mining activities. Ambient noise analysis and imaging can be cost-effective but will require more time before it sees routine use and garners the trust necessary for risk analysis. More tests and extensive analyses are still required.

Recently, seismological array data gathered near a river in Switzerland were used by Roth et al. (2016) to characterize the sediment transport with respect to discharge and precipitation. Decomposition of spectral data was intended to provide a method for bed load monitoring, expressing the growing adaptation of seismological methods for near-surface research.

4.2 Advanced and Cost-Effective Airborne Measurements

Recent developments in AEM data acquisition systems (e.g., Auken et al., 2009; Sørensen & Auken, 2004) have allowed acquisition of airborne TEM measurements that possess high accuracy and resolution for mapping near surface structures. Anschütz, Christensen, and Pfaffhuber (2014) present a successful example where airborne measurements were integral to safely and cost-effectively completing a new major highway between eastern Norway and Sweden. To estimate the depth to bedrock and map potential areas with sensitive clays, an AEM survey using the SkyTEM 302 system was acquired along a planned 32-km-long motorway. Acquisition took 3 days and resulted in coverage along 178 km of flight lines. Bedrock topography is clearly evident in the resistivity models along the flight lines. Modeling results also revealed areas with potentially sensitive clay deposits. Good correlation was observed between resistivity models and geotechnical boreholes. A 600-m-long ERT profile was then acquired along one of the AEM flight lines and near the city of Oslo where a known quick-clay zone existed to better assess the performance of the airborne TEM method (Anschütz et al., 2016) (Fig. 36). Both a constrained (by bedrock elevation) and unconstrained ERT inversions were performed. The AEM model exposes internal structures in the clay layer. The unconstrained ERT model possesses a smooth transition from conductive clay to resistive rock due to the inversion process. Constrained ERT model better depicts bedrock and

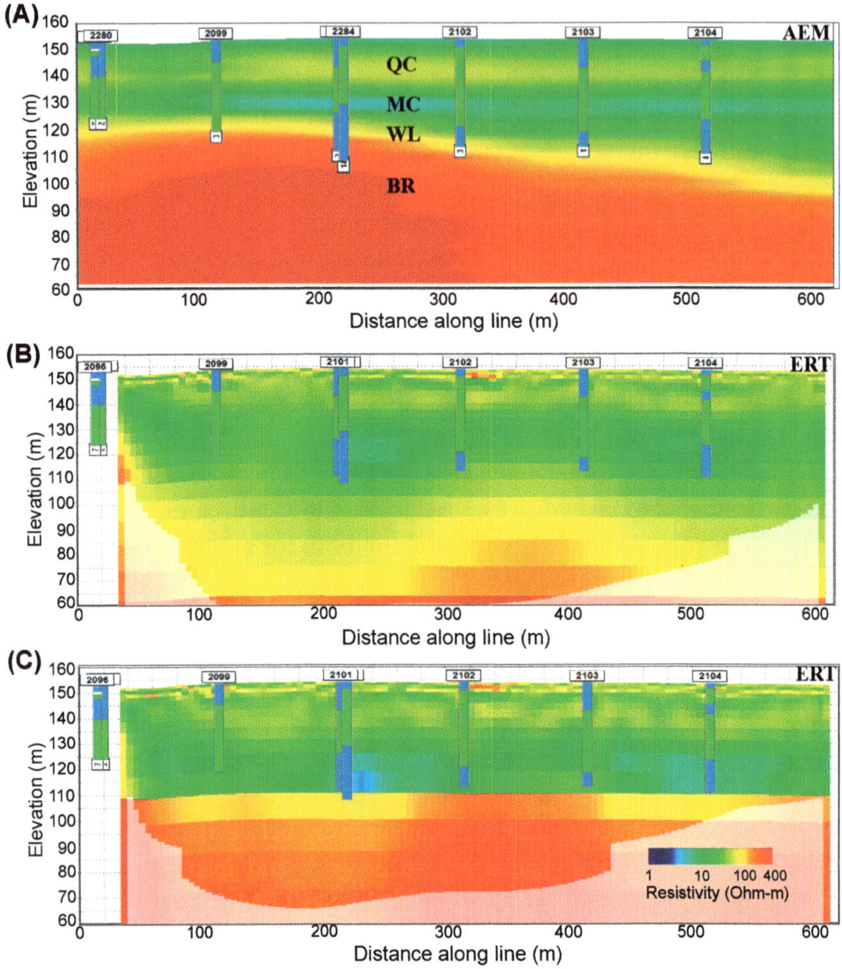

Figure 36 (A) Example of an airborne time-domain electromagnetic model, (B) unconstrained electrical resistivity tomography (ERT) model, and (C) constrained by bedrock depth from boreholes ERT model. The letters in the top panel mark the four main layers: possible quick clay (QC), marine clay (MC), weathered layer (WL), and bedrock (BR). Quick clay identified with coring is marked in green. *Modified from Anschütz, H., Pfaffhuber, A.A., Domaas, U., & Rosenvold, B.St. (2015). Combined airborne and ground geophysics as a first phase towards a landslide warning system — a Norwegian case study.* Österreichische Ingenieur- und Architekten-Zeitschrift, *160. (Jg).*

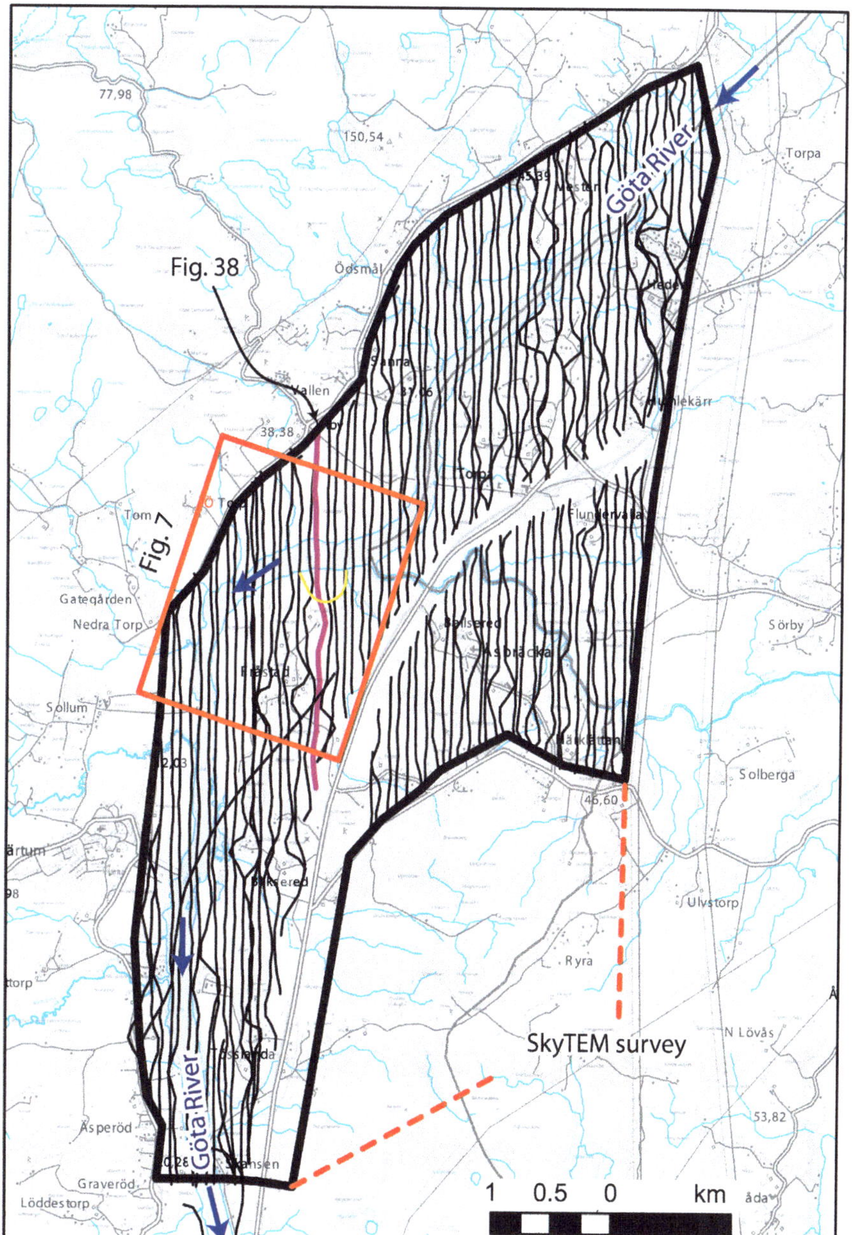

Figure 37 Airborne time-domain electromagnetic flight lines covering the quick-clay landslide test site shown in Fig. 7. The area was chosen to test the applicability of airborne electromagnetic method for quick-clay mapping and in particular due to existing ground-based geotechnical (e.g., Löfroth et al., 2011; Salas-Romero et al., 2016) and geophysical (e.g., Malehmir, Bastani et al., 2013) measurements available from the site. *Purple line* shows location of a resistivity model derived from the data and shown in Fig. 38A.

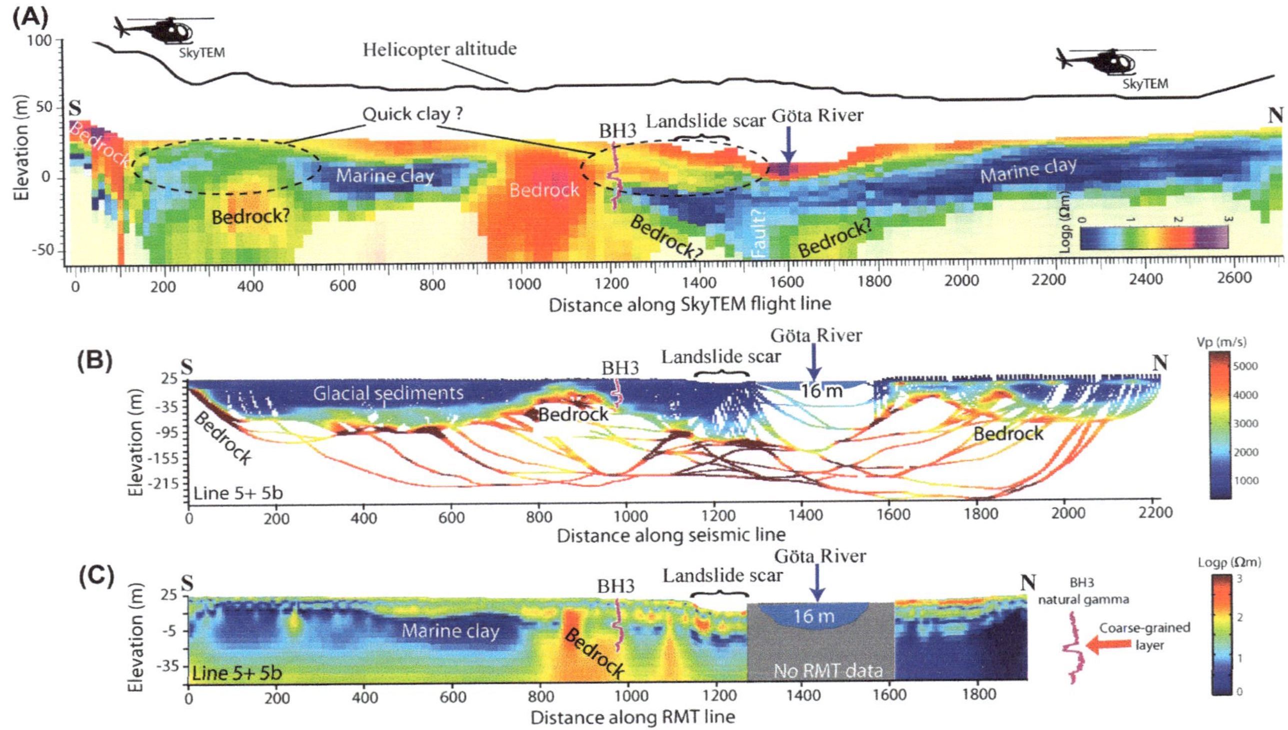

Figure 38 (A) Resistivity cross section derived from airborne time-domain electromagnetic data (Fig. 37) and compared with (B) seismic tomography (modified from Wang et al., 2016), and (C) ground-based radio magnetotelluric results *(Modified from Wang, S., Malehmir, A., & Bastani, M. (2016). Geophysical characterization of areas prone to quick-clay landslides in southwestern Sweden using radio-magnetotelluric and seismic methods. Tectonophysics, 677–678, 248–260.)*. Decrease in the natural gamma data is associated with a coarse-grained layer acted as a reflector as shown in Fig. 7. See text for details.

pockets of marine clay over the bedrock ($\rho < 10\ \Omega\text{m}$) at depth. Quick-clay materials are evident on the top of the models as zones of elevated resistivity.

A quick-clay test site in Sweden (Löfroth et al., 2011; Malehmir, Bastani et al., 2013) along with several other locations with recently acquired airborne TEM measurements were the focus of a study comparing different geophysical methods to detect and image quick clay. The Swedish Civil Contingencies Agency (MSB) supported this study. A total of 850 km of flight line data were collected during 5 days (Fig. 37 and Fig. 7) with a nominal flight line separation of 75 m and altitude of 50 m. The resistivity cross section derived from the airborne data along one of the seismic profiles (profile 5 + 5b in Fig. 7) is of good quality (Fig. 38). The AEM data were inverted using the SCI algorithm developed by Viezzoli et al. (2008). Seismic and RMT data are also available from the same location (Wang et al. 2016) allowing for direct comparisons (Fig. 38).

Comparing the three models reveals that the deepest penetration was obtained by the refraction seismic method. It resolved bedrock surface (Fig. 38B), but it showed no sensitivity to some of the near-surface materials resolved by the ERT and AEM methods. While bedrock is best resolved in the tomographic section, AEM data provide greater details in the normally consolidated sediments as well as evidence of the presence of a fault near the river (this was speculated in the earlier studies too, e.g., Adamczyk, Malinowski, & Malehmir, 2014; Lundberg et al., 2014). Marine clays dominate on the northern side of the river while on the southern side, leached clays can be inferred consistent with high resistivities.

Ground-RMT data are also consistent with the AEM data and show much higher-resolution close to the surface and lack much detail at deeper levels. Interestingly, the downhole natural gamma data correlate well with the resistive layer imaged in the AEM data, which is a combination of coarse-grained (mainly) and quick-clay materials. These examples illustrate the potential of airborne geophysical methods as a cost-effective mapping tool for areas prone to slides and highlight the need for integrated approaches.

4.3 Advanced and Cost-Effective 2D and 3D DCIP Measurements

With the advancements in ERT methods cost-effective data are now routinely acquired by many organizations. Several commercial ERT instruments currently available can also acquire time-domain IP data, which combined with resistivity allows acquisition of so-called DCIP data. IP data are usually inverted to calculate integral apparent chargeability, which can lead to different outcomes depending on the selected transmitted current pulse and integration interval if not accounted for by the inversion algorithm (e.g., Olsson, Dahlin, et al., 2015; Olsson, Fiandaca, et al., 2015). Shortening time measurements below what is optimal for DCIP data acquisition is tempting to save time and reduce cost, but this compromise decreases the signal-to-noise ratio and reduces the spectral content of the data.

To improve the situation, recording waveform data and the measurements of IP response during on-time can be done and this will lead to a slight increase in the signal quality (Olsson, Dahlin, et al., 2015; Olsson, Fiandaca, et al., 2015). Extraction of spectral IP (SIP) parameters from DCIP data (Fiandaca et al., 2013) opens up new possibilities to better-determined chargeability as well as other Cole—Cole parameters. The advantage of time-domain SIP over frequency-domain SIP is the faster and more efficient field data acquisition possible using ERT-style multielectrode and multichannel instruments. Limitations include a much higher sensitivity to noise requiring careful attention to field data acquisition procedures, and the need for an extended time for recording longer IP time windows with acceptable data quality. The latter can be addressed with data processing and filtering tools that have been shown to double the effective spectral range (Olsson, Fiandaca, Juul Larsen, Dahlin, & Auken, 2016).

3D ERT has significant potential as a tool for characterizing complex geological environments, with the 3D aspect often a necessity for geological mapping of geohazards. 3D ERT is, however, not yet a standard and fully accepted tool industry wide, with the combined process of survey design, data acquisition, data quality assessment, data processing, and inversion often involving a lot of time resources. Data acquisition can be time-consuming considering the large amount of data acquired for the method to be effective, and with the clear need for optimized and time-efficient methodology and instrumentation for 3D geoelectrical surveying, especially if it is going to measure DCIP data. Some approaches for optimized design of measurement sequences have been presented (e.g., Blome, Maurer, & Greenhalgh, 2011; Loke, Wilkinson, Uhlemann, Chambers, & Oxby, 2014), but the viability of the approaches is intimately coupled to the flexibility and limitations of the hardware design.

ERT monitoring (time-lapse) is an emerging technology with a large potential as demonstrated in a number of recent applications for different types of geohazards (Gruber, 2015; Gruber, Jihyang, & Minsun, 2013; Supper & Kauer, 2011) including landslide risk (Supper et al., 2014;

Uhleman et al., 2015), permafrost thawing (Hilbich et al. 2008), detection of internal erosion, and anomalous seepage (Johansson & Dahlin, 1996; Sjödahl, Dahlin, & Johansson, 2009). Several of the commercially available data acquisition systems can be configured for automatic monitoring but they are generally not optimized for it in terms of cost-performance and flexibility of system setup. Institutions such as geological surveys have developed in-house systems specifically designed for long-term monitoring. Longer-term monitoring generates large quantities of data, with data quality assessment, inversion processing, visualization, and archiving of these data presenting unique challenges that must be resolved for routine applications. Monitoring with DCIP is an expansion that would add further nuances to the output, however, it would come with further demands on data handling and storing especially if full waveform data are saved.

4.4 A Complex Problem Requiring a Multidisciplinary and Integrated Approach

Comprehending the mechanisms and settings that lead to geohazards is the key starting point for risk analysis and threat mitigation. It is the complexity of subsurface models that complicates and dictates the distribution of different physical parameters related to structural framework, mechanical properties, and hydrogeological regimes. More often these complexities are better studied when integrating several geophysical techniques. Data integration has a couple advantages: one relates to a more comprehensive characterization of the subsurface, and the other is the independent validation of each survey's results. Many geophysical methods rely on indirect techniques based on processing and inversion, and therefore suffer from solution nonuniqueness and method-specific intrinsic limitations. Hence the integration of different techniques can provide an overall result with a higher reliability and fidelity (Garofalo, 2014 and reference therein).

There are different approaches for integrating different methods and data. These range from *a posteriori* comparisons of independent results to sophisticated joint inversion algorithms where different kinds of data sets are inverted jointly by imposing structural and sometimes physical compliance to the same multiparametric model (e.g., Socco, Foti, et al., 2010 and references therein). Many different approaches are presented in the literature for joint inversion of geophysical data and attention to the potential of joint inversion has grown significantly in recent years (SEG digital library, for instance, contains twice the documents, 214, about joint inversion from 2010 to present, than those, 107, produced in the previous 80 years).

Several examples presented in this paper make use of data integration at various levels (see examples in Figs. 10, 13 and 29) with two examples (Figs. 11 and 32) dealing with joint inversion of different types of data. In the latter example, the joint inversion of seismic surface and body wave data is performed by imposing the same structure and a physically consistent value of Poisson's ratio. The results provided here can be compared with individual inversion results obtained by Konstantaki et al. (2013) on the same data set. There is general agreement between the two results but as would be expected individual inversion schemes provided some local inconsistencies, which joint inversion overcomes. Another example of joint inversion applied to geohazard studies can be found in Garofalo, Sauvin, Socco, and Lecomte (2015), where seismic and resistivity data were jointly inverted to delineate a quick-clay deposit. It was evident that better agreement was obtained between the joint inversion results and existing borehole data than the individual inversions. The inclusion of petrophysical relationships can further improve the capability of joint inversion results to represent reliable and comprehensive physical models of the subsurface and it is certainly an advancement in our science with great potential for routine incorporation into geohazard studies in the foreseeable future.

5. CONCLUSIONS

Geophysical methods have proven to possess great potential for detailed characterization and delineation of subsurface geology in settings with geohazards that may lead to disastrous failures when triggered by any of a variety of sources such as heavy rain, earthquake, volcanic activity, and erosion. We have presented a number of methods and their underlying principles and limitations typically used for studying geohazards and helping to design programs for mitigating risks associated with them. While examples provided here are biased towards developed countries, they illustrate what can be achieved using these methods when properly planned, acquired, processed, and implemented within the limits of the methods. Airborne geophysical methods will continue to be dominant for fast, cost-effective, and large-scale studies (e.g., nation-wide programs). Ground-based methods will be applied to very specific problems or studies requiring great detail. While surficial investigation will be used for a first hand hazard assessment, it is the underlying geology that controls the risks associated with a particular geohazard. We have briefly reviewed potential areas for improvements and where improvement is on the horizon, particularly as it relates to fast and cost-effective data acquisition and improved subsurface imaging and interpretations. It is the hope of the authors that by showing the utility of integrating various geophysical methods with relevant geotechnical, geomechanical, and hydrogeological data applied to geohazards that potentially threaten the environmental and/or infrastructure, various national and international programs will encourage and support the use of geophysics in geohazard characterization, thereby reducing the overall cost of characterization and reduce the risk to public safety.

ACKNOWLEDGMENTS

This article was initiated through various projects particularly sponsored by the Geoscientists *Without* Borders (GWB) program of the Society of Exploration Geophysicists (SEG) for which we are grateful. A. Malehmir thanks the Trust 2.2 project (Trust-GeoInfra: http://www.trust-geoinfra.se) and its sponsors: Formas (252-2012-1907), SGU, BeFo, SBUF, Skanska, NGI, and FQM through which some of the works presented in this article and collaboration among some of the authors were initiated. We thank the editor L. Nielsen for useful and critical comments that improved the overall quality of the paper. We thank Nelson F. Fernnandes for providing the photos from the 2011 Rio de Janeiro landslide. We thank Mark Reid for providing the photo from the 2014 Oso landslide. We thank Helgard Anschütz for sharing some of the figures shown in this paper from the Norwegian quick-clay site. GMT from P. Wessel and W.H.F. Smith and GOCAD were used to prepare some of the figures.

REFERENCES

Adamczyk, A., Malinowski, M., & Malehmir, A. (2013). Application of first-arrival tomography to characterize a quick-clay landslide site in southwest Sweden. *Acta Geophysica, 61*, 1057–1073.

Adamczyk, A., Malinowski, M., & Malehmir, A. (2014). High-resolution near-surface velocity model building using full-waveform inversion—a case study from southwest Sweden. *Geophysical Journal International, 197*, 1693–1704.

Albarello, D., Socco, L. V., Picozzi, M., & Foti, S. (2015). Seismic hazard and land management policies in Italy: the role of seismic investigation. *First Break, 33*, 87–93.

Anschütz, H., Bazin, S., Kåsin, K., Pfaffhuber, A. A., & Smaavik, T. F. (2016). Airborne sensitive clay mapping – stretching the limits of AEM resolution and accuracy. *Near Surface Geophysics* (submitted).

Anschütz, H., Christensen, C., & Pfaffhuber, A. A. (2014). *Quantitative depth to bedrock extraction from AEM data, extended abstracts, 20th European meeting of environmental and engineering geophysics (EAGE's near surface 2014), Athens, Greece*. http://dx.doi.org/10.3997/2214-4609.20141994.

Anschütz, H., Pfaffhuber, A. A., Domaas, U., & Rosenvold, B. St. (2015). Combined airborne and ground geophysics as a first phase towards a landslide warning system – a Norwegian case study. *Österreichische Ingenieur- und Architekten-Zeitschrift, 160* (Jg).

Armijo, R., Pondard, N., Meyer, B., Ucarkus, G., Mercier de Lépinay, B., & the MARMARASCARPS cruise party. (2005). Submarine fault scarps in the Sea of Marmara pull-apart (North Anatolian Fault): Implications for seismic hazard in Istanbul. *Geochemistry, Geophysics Geosystems, 6*, 1–29.

Arts, R., Chadwick, A., Eiken, O., Thibeau, S., & Nooner, S. (2008). Ten years' experience of monitoring CO_2 injection in the Utsira Sand at Sleipner, offshore Norway. *First Break, 26*, 65–72.

Auken, E., & Christiansen, A. V. (2004). Layered and laterally constrained 2D inversion of resistivity data. *Geophysics, 69*, 752–761.

Auken, E., Christiansen, A. V., Westergaard, J. A., Kirkegaard, C., Foged, N., & Viezzoli, A. (2009). An integrated processing scheme for high-resolution ailocatrborne electromagnetic surveys, the SkyTEM system. *Exploration Geophysics, 40*, 184–192.

Auken, E., Pellerin, L., Christensen, N. B., & Sørensen, K. I. (2006). A survey of current trends in near-surface electrical and electromagnetic methods. *Geophysics, 71*, G249–G260.

Baldwin, W. E., Morton, R. A., Denny, J. F., Dadisman, S. V., Schwab, W. C., Gayes, P. T., & Driscoll, N. W. (2004). *Maps showing the stratigraphic framework of South Carolina's long bay from Little River to Winyah Bay. U.S. Geological survey*. Open File Report 2001-1013.

Bamler, R., & Hartl, P. (1998). Synthetic aperture radar interferometry. *Inverse Problems, 14*, R1–R54.

Bardi, F., Raspini, F., Ciampalini, A., Kristensen, L., Rouyet, L., Lauknes, T., ... Casagli, N. (2016). Space-borne and ground-based InSAR data integration: the åknes test site. *Remote Sensing, 8*(3), 237. http://dx.doi.org/10.3390/rs8030237.

deBari, C., Lapenna, V., Perrone, A., Puglisi, C., & Sado, F. (2011). Digital photogrammetric analysis and electrical resistivity tomography for investigating the Picerno landslide (Basilicata region, southern Italy). *Geomorphology, 133*, 34–46.

Bastani, M. (2001). *EnviroMT — A new controlled source/radio magnetotelluric system* (Ph.D. thesis). Uppsala University.

Bastani, M., Hübert, J., Kalscheuer, T., Pedersen, L. B., Godio, A., & Bernard, J. (2012). 2D joint inversion of RMTand ERT data versus individual 3D inversion of full tensor RMT data: an example from Trecate site in Italy. *Geophysics*, 77, WB233–WB243.

Bastani, M., Persson, L., Mehta, S., & Malehmir, A. (2015). Boat-towed radiomagnetotellurics (RMT) — A new technique and case study from the city of Stockholm. *Geophysics, 80*, B193–B202.

Bastani, M., Savvaidis, A., Pedersen, L. B., & Kalscheuer, T. (2011). CSRMT measurements in the frequency range of 1–250 kHz to map a normal fault in the Volvi basin, Greece. *Journal of Applied Geophysics, 75*, 180–195.

Bazin, S., Feuillet, N., Duclos, C., Crawford, W., Nercessian, A., Bengoubou-Valerius, M., ... Singh, S. C. (2010). The 2004–2005 Les Saintes (French West Indies) seismic aftershock sequence observed with ocean bottom seismometers. *Tectonophysics, 489*, 1–4.

Bazin, S., & Pfaffhuber, A. (2013). Mapping of quick clay by electrical resistivity tomography under structural constraint. *Journal of Applied Geophysics, 98*, 280–287.

Beck, B. F., & Sinclair, W. C. (1986). *Sinkholes in Florida: An introduction*. The Florida Sinkhole Research Institute at the University of Central Florida.

Beilecke, T., Krawczyk, C. M., Ziesch, J., & Tanner, D. C. (2016). Near-surface fault detection using high resolution shear wave reflection seismics at the CO2CRC Otway Project

Site. *Journal of Geophysical Research, Solid Earth, 121*. http://dx.doi.org/10.1002/2015JB012668.

Benito, B., Cepeda, J. M., & Martinez Diaz, J. J. (2004). Analysis of the spatial and temporal distribution of the 2001 earthquakes in El Salvador. In W. I. Rose, J. J. Bommer, D. L. Lopez, M. J. Carr, & J. J. Major (Eds.), *Geological Society of America Special Paper 375, Natural Hazards in El Salvador* (pp. 339–356). Boulder, Colorado.

Bergamo, P., Boiero, D., & Socco, L. V. (2012). Retrieving 2D structures from surface wave data by means of a space-varying spatial windowing. *Geophysics, 77*, EN39–EN51.

Bergamo, P., & Socco, L. V. (2014). Detection of sharp lateral discontinuities through the analysis of surface-wave propagation. *Geophysics, 79*(4), EN77–EN90.

Bichler, A., Bobrowsky, P., Best, M., Douma, M., Hunter, J., Calvert, T., & Burns, R. (2004). Three dimensional mapping of a landslide using a multi-geophysical approach: the Quesnel Forks landslide. *Landslides, 1*, 29–40.

Binley, A., Hubbard, S. S., Huisman, J. A., Revil, A., Robinson, D. A., Singha, K., & Slater, L. D. (2015). The emergence of hydrogeophysics for improved understanding of subsurface processes over multiple scales. *Water Resources Research, 51*, 3837–3866. http://dx.doi.org/10.1002/2015WR017016.

Bjerrum, L. (1954). Geotechnical properties of Norwegian marine clays. *Géotechnique, 4*, 49–69.

Blikra, L. H. (2012). The åknes rockslide, Norway. In J. J. Clague, & D. Stead (Eds.), *Landslides: Types, mechanisms and modeling* (pp. 323–335). Cambridge, UK: Cambridge University Press.

Blindow, N., Salat, C., Gundelach, V., Buschmann, U., & Kahnt, W. (2011). *Performance and calibration of the helicoper GPR system BGR-P30. Advanced ground penetrating radar (IWAGPR), 2011 6th International Workshop.*

Blome, M., Maurer, H., & Greenhalgh, S. (2011). Geoelectric experimental design — efficient acquisition and exploitation of complete pole-bipole data sets. *Geophysics, 76*, F15–F26.

Boaga, J., Vignoli, G., & Cassiani, G. (2011). Shear wave profiles from surface wave inversion: the impact of uncertainty on seismic site response analysis. *Journal of Geophysics and Engineering, 8*, 162–174.

Bogoslovsky, V. A., & Ogilvy, A. A. (1977). Geophysical methods for the investigation of landslides. *Geophysics, 42*, 562–571.

Boiero, D., & Socco, L. V. (2014). Joint inversion of Rayleigh-wave dispersion and P-wave refraction data for laterally varying layered models. *Geophysics, 79*, EN49–EN59.

Bostick, F. (2000). Field experimental results of three-component fiber-optic seismic sensors. In *65th Annual International Meeting*. SEG. http://dx.doi.org/10.1190/1.1815889. Expanded Abstracts.

Bottelin, P., Jongmans, D., Baillet, L., Lebourg, T., Hantz, D., Lvy, C., ... Darras, L. (2013). Spectral Analysis of Prone-to-fall Rock Compartments using Ambient Vibrations. *Journal of Environmental & Engineering Geophysics, 18*(4), 205–217. http://dx.doi.org/10.2113/JEEG18.4.205.

Bour, M., Fouissac, D., Dominique, P., & Martin, C. (1998). On the use of microtremor recordings in seismic microzonation. *Soil Dynamics and Earthquake Engineering, 17*(7–8), 465–474. http://dx.doi.org/10.1016/S0267-7261(98)00014-1.

Brodic, B., Malehmir, A., Juhlin, C., Dynesius, L., Bastani, M., & Palm, H. (2015). Multicomponent broadband digital-based seismic landstreamer for near surface applications. *Journal of Applied Geophysics, 123*, 227–241.

Brothelande, E., Finizola, A., Peltier, A., Delcher, E., Komorowski, J.-C., Di Gangi, F., ... Legendre, Y. (2014). Fluid circulation pattern inside La Soufrière volcano (Guadeloupe) inferred from combined electrical resistivity tomography, self-potential, soil temperature and diffuse degassing measurements. *Journal of Volcanology and Geothermal Research, 288*, 105–122.

Brown, L. T., Diehl, J. G., & Nigbor, R. L. (2000). A simplified procedure to measure average shear-wave velocity to a depth of 30 meters (VS30). In *12th World Conference on Earthquake Engineering*.

Buchli, T., Merz, K., Zhou, X. H., Kinzelbach, W., & Springman, S. M. (2013). Characterization and monitoring of the Furggwanghorn Rock Glacier, Turtmann Valley, Switzerland: results from 2010 to 2012. *Vadose Zone Journal, 12*(1). http://dx.doi.org/10.2136/vzj2012.0067.

Burger, H. R., Sheehan, A. F., & Jones, C. H. (2006). *Introduction to applied geophysics: Exploring the shallow subsurface*. New York: W.W. Norton, ISBN 9780393926378, 600 pp.

Burschil, T., Beilecke, T., & Krawczyk, C. M. (2015). Finite difference modelling to evaluate seismic P-wave and shear-wave field data. *Solid Earth, 6*, 33–47.

Burschil, T., & Krawczyk, C. M. (2016). How fractures mask a cavity for shear waves. *Geophysical Journal International* (submitted).

Butler, D. K. (1984). Microgravimetric and gravity gradient techniques for detection of subsurface cavities. *Geophysics, 49*, 1084–1096.

Bye, B. L. (2011). *Volcanic eruptions: Science and risk management*. http://www.science20.com/planetbye/volcanic_eruptions_science_and_risk_management-79456.

Calvert, A. J., Fisher, M. A., & Ships working group. (2001). Imaging the Seattle Fault Zone with high-resolution seismic tomography. *Geophysical Research Letters, 28*, 2337–2340.

Carpentier, S. F. A., Green, A. G., Horstmeyer, H., Kaiser, A. E., Hurter, F., Langridge, R., & Finnemore, M. (2010). *Reflection seismic surveying across the Alpine Fault immediately north of the intersection with the Hope Fault*. Near Surface 2010. Zürich, Extended Abstract B22.

Carpentier, S. F. A., Green, A. G., Langridge, R., Boschetti, S., Doetsch, J. A., Abaecherli, A., et al.Finnemore, M. (2012). Flower structures and Riedel shears at a step-over zone along the Alpine Fault (New Zealand) inferred from 2D and 3D GPR images. *Journal of Geophysical Research, 117*, B02406.

Carroll, R. D., Scott, J. H., & Lee, F. T. (1972). Seismic refraction studies, in Geological, geophysical and engineering investigations of the Loveland Basin landslide, Clear Creek County, Colorado, 1963–65. *U.S. Geological Survey Professional Paper 673 C*, 17–19.

Cassiani, G., Ursino, N., Deiana, R., Vignoli, G., Boaga, J., Rossi, M., … Werban, U. (2012). Noninvasive monitoring of soil static characteristics and dynamic states: a case study highlighting vegetation effects on agricultural land. *Vadose Zone Journal, 11*. http://dx.doi.org/10.2136/Vzj2011.0195.

Cauzzi, C., Faccioli, E., & Costa, G. (2011). 1D and 2D site amplification effects at Tarcento (Friuli, NE Italy), 30 years later. *Journal of Seismology, 15*, 1–17. http://dx.doi.org/10.1007/s10950-010-9202-y.

Chambers, J. E., Meldrum, P. I., Gunn, D. A., Wilkinson, P. B., Kuras, O., Weller, A. L., & Ogilvy, R. D. (2009). Hydrogeophysical monitoring of landslide processes using automated time-lapse electrical resistivity tomography (ALERT). Near surface geoscience. In *15th European meeting of environmental and engineering geophysics Dublin, Ireland, 7–9, September 2009*.

Chen, H.-W., & McMechan, G. A. (1993). 3-D physical modeling and pseudospectral simulation of seismic common-source data volumes. *Geophysics, 58*, 121–133.

Cheraghi, S., Malehmir, A., & Bellefleur, G. (2012). 3D imaging challenges in steeply dipping mining environment: new lights on acquisition geometry and processing from the Brunswick No. 6 seismic data, Canada. *Geophysics, 77*, WC109–WC122.

Cheraghi, S., Malehmir, A., Bellefleur, G., Bongajum, E., & Bastani, M. (2013). Scaling behavior and the effects of heterogeneity on shallow seismic imaging of mineral deposits: a case study from Brunswick No. 6 mining area, Canada. *Journal of Applied Geophysics, 90*, 1–18.

Chouet, B. A., Dawson, P., Ohminato, T., Martini, M., Saccorotti, G., Giudicepietro, F., ... Scarpa, R. (2003). Source mechanism of explosions at stromboli determined from moment tensor inversions of very-long-period data. *J. Geophys. Res, 108*(B1), 2019. http://dx.doi.org/10.1029/2002JB001919.

Claprood, M., & Asten, M. W. (2009). Initial results from spatially averaged coherency, frequency-wavenumber, and horizontal to vertical spectrum ratio microtremor survey methods for site hazard study at Launceston, Tasmania. *Exploration Geophysics, 40*(1), 132–142.

Cluff, L. S. (1971). Peru earthquake of May 31, 1970: engineering geology observations. *Bulletin of the Seismological Society of America, 61*, 511–533.

Comina, C., & Foti, S. (2015). Comment on "Implications of surface wave data measurement uncertainty on seismic ground response analysis" by Jakka et al. *Soil Dynamics and Earthquake Engineering, 74*, 89–91.

Comina, C., Foti, S., Boiero, D., & Socco, L. V. (2011). Reliability of VS,30 evaluation from surface waves tests. *Journal of Geotechnical and Geoenvironmental Engineering, 137*, 579–588.

Cornou, C., Ohrnberger, M., Boore, D. M., Kudo, K., & Bard, P.-Y. (2009). *Using ambient noise array techniques for site characterization: results from an international benchmark* (pp. 30–92). Grenoble: Geology on Seismic Motion.

Cosentino, P., Martorana, R., Perniciaro, M., & Terranova, L. M. (2003). Geophysical study of an landslide in northern Sicily. *Near Surface Geophysics, 1*, 77–84.

COST 65. (1995). Hydrogeological aspects of groundwater protection in karstic areas. In *Final Report Europ. Comm., Report EUR16547, 446 pp., Brussels.*

Cox, L. H., Wilson, G. A., & Zhdanov, S. (2012). 3D inversion of airborne electromagnetic data. *Geophysics, 77*, 59–69.

Cox, B. R., Wood, C. M., & Teague, D. P. (2014). Synthesis of the UTexas1 surface wave dataset blind-analysis study: interanalyst dispersion and shear wave velocity uncertainty. In *Geo-congrees. Atlanta Georgia.*

Cruden, D. M., & Varnes, D. J. (1996). Landslide types and processes, in landslides. In A. K. Turner, & R. L. Schuster (Eds.), *Transport research board* (pp. 36–75). Washington DC: National Research Council.

Dahlin, T. (2001). The development of electrical imaging techniques. *Computers and Geosciences, 27*, 1019–1029.

Dahlin, T., & Leroux, V. (2012). Improvement in time-domain induced polarisation data quality with multi-electrode systems by separating current and potential cables. *Near Surface Geophysics, 10*, 545–565.

Dahlin, T., Löfroth, H., Schälin, D., & Seur, P. (2013). Mapping of quick clay using geoelectrical imaging and CPTU-resistivity. *Near Surface Geophysics, 11*, 659–670.

Dahlin, T., & Zhou, B. (2006). Multiple-gradient array measurements for multichannel 2D resistivity imaging. *Near Surface Geophysics, 4*, 113–123.

Dahm, T., Heimann, S., & Bialowons, W. (2011). A seismological study of shallow weak micro-earthquakes in the urban area of Hamburg city, Germany, and its possible relation to salt dissolution. *Natural Hazards*. http://dx.doi.org/10.1007/s11069-011-9716-9.

Dahm, T., Kühn, D., Ohrnberger, M., Kröger, J., Wiederhold, H., Reuther, C., ... Scherbaum, F. (2010). Combining geophysical data sets to study the dynamics of shallow evaporates in urban environments. *Geophysical Journal International, 181*. http://dx.doi.org/10.111/j.365-246X.2010.04521.x.

Daley, B., Freifeld, M., Ajo-Franklin, J., Dou, S., Pevzner, R., Shulakova, V., ... Lueth, S. (2013). Field testing of fiber-optic distributed acoustic sensing (DAS) for subsurface seismic monitoring. *The Leading Edge, 32*, 699–706. http://dx.doi.org/10.1190/tle32060699.1.

Daniell, J., Wenzel, F., & Schaefer, A. (2016). The economic costs of natural disasters globally from 1900–2015: historical and normalised floods, storms, earthquakes, volcanoes, bushfires, drought and other disasters. *Geophysical Research Abstracts, 18*. EGU2016-1899, 2016 EGU General Assembly 2016 http://meetingorganizer.copernicus.org/EGU2016/EGU2016-1899.pdf.

Davis, K., Li, Y., & Batzle, M. (2008). Time-lapse gravity monitoring: a systematic 4D approach with application to aquifer storage and recovery. *Geophysics, 73*, WA61–WA69.

Dehghannejad, M., Bauer, T., Malehmir, A., Juhlin, C., & Weihed, P. (2012). Crustal geometry of the central Skellefte district, northern Sweden-constraints from reflection seismic investigations. *Tectonophysics, 524-525*, 87–99.

Dentith, M., & Mudge, S. T. (2014). *Geophysics for the mineral exploration geoscientist. 9780521809511*. Cambridge University Press, 454 pp.

Dobecki, T. L. (2010). Sinkholes and pitfalls in urban geophysics. *The Leading Edge, 29*, 944–951.

Doneus, M., Miholjek, I., Mandlburger, G., Doneus, N., Verhoeven, G., Briese, C. H., & Pregesbauer, M. (2015). Airborne laser bathymetry for documentation of submerged archaeological sites in shallow water. *ISPRS – International Archives of the Photogrammetry, Remote Sensing and Spatial Information Sciences*, 99–107. http://dx.doi.org/10.5194/isprsarchives-xl-5-w5-99-2015. XL-5/W5.

Donohue, S., Long, M., O'Connor, P., Helle, T. E., Pfaffhuber, A., & Rømoen, M. (2012). Multi-method geophysical mapping of quick clay. *Near Surface Geophysics, 10*, 207–219.

Dumont, M., Reninger, P. A., Pryet, A., Martelet, G., Aunay, B., & Join, J. L. (2015). A statistical approach to interpret regional airborne electromagnetic surveys for hydrogeological characterization. In *EAGE Near Surface Geoscience Conference, Torino*.

D'Alessandro, A., & Martorana, R. (2015). *Drones: New technologies for geophysics? Near surface geoscience, 2015, Turin, Italy*. http://dx.doi.org/10.3997/2214-4609.201413681.

Eckerstorfer, M., Bühler, Y., Frauenfelder, R., & Malnes, E. (2016). Remote sensing of snow avalanches: recent advances, potential, and limitations – a review. *Cold Regions Science and Technology, 121*, 126–140. http://dx.doi.org/10.1016/j.coldregions.2015.11.001.

Eidsvig, U. M., Medina-Cetina, Z., Kveldsvik, V., Glimsdal, S., Harbitz, C. B., & Sandersen, F. (2009). Risk assessment of a tsunamigenic rockslide at Åknes. *Natural Hazards, 56*, 529–545.

Eppelbaum, L. V., Ezersky, M., Al-Zoubi, A., Goldshmidt, V., & Legchenko, A. (2008). Study of the factors affecting the karst volume assessment in the Dead Sea sinkhole problem using microgravity field analysis and 3-D modeling. *Advances in Geosciences, 19*, 97–115.

Eurocode 8, EC8. (2003). Draft prENV 1998-1 Jan 2003, design of structures for earthquakes resistance, Part 1: general rules, seismic actions and rules for buildings (Draft No 4). In *European committee for standardization, Brussels Belgium*.

Everett, M. E. (2013). *Near-surface applied geophysics*. http://dx.doi.org/10.1017/CBO9781139088435. ISBN:10:1107018773.

Ezersky, M. G., Eppelbaum, L. V., Al-Zoubi, A., Keydar, S., Abueladas, A., Akkawi, E., & Medvedev, B. (2013). Geophysical prediction and following development sinkholes in two Dead Sea areas, Israel and Jordan. *Environmental Earth Sciences, 70*, 1463–1478. http://dx.doi.org/10.1007/s12665-013-2233-2.

Ferrigno, F., Gigli, G., Fanti, R., & Casagli, N. (2015). GB-InSAR monitoring and observational method for landslide emergency management: the Montaguto earthflow (AV, Italy). *Natural Hazards and Earth System Sciences Discuss., 3*, 7247–7273. http://dx.doi.org/10.5194/nhessd-3-7247-2015, 201.

Fiandaca, G., Ramm, J., Binley, A., Gazoty, A., Christiansen, A. V., & Auken, E. (2013). Resolving spectral information from time domain induced polarization data through 2-D inversion. *Geophysical Journal International, 192*, 631–646.

Finn, C. A., Sisson, T. W., & Deszcz-Pan, M. (2001). Aerogeophysical measurements of collapse-prone hydrothermally altered zones at Mount Rainier volcano. *Nature, 409*, 600–603.

Foti, S., Comina, C., Boiero, D., & Socco, L. V. (2009). Non uniqueness in surface wave inversion and consequences on seismic site response analyses. *Soil Dynamics and Earthquake Engineering, 29*, 982–983.

Foti, S., Lai, C. G., Rix, G. J., & Strobbia, C. (2014). *Surface wave methods for near-surface site characterization*. Boca Raton, Florida (USA): CRC Press, ISBN 9780415678766, 487 pp.

Foti, S., Parolai, S., Albarello, D., & Picozzi, M. (2011). Application of surface wave methods for seismic site characterization. *Survey in Geophysics, 32*, 777–825.

Fountain, D. (1998). Airborne electromagnetic systems – 50 years of development. *Exploration Geophysics, 29*, 1–11.

Garofalo, F. (2014). *Physically constrained joint inversion of seismic and electrical data for near surface applications* (Ph.D. dissertation). Politecnico di Torino.

Garofalo, F., Foti, S., Hollender, F., Bard, P. Y., Cornou, C., Cox, B. R., … Vergniault, C. (2016b). InterPACIFIC project: comparison of invasive and non-invasive methods for seismic site characterization. Part II: Inter-comparison between surface-wave and borehole methods. *Soil Dynamics and Earthquake Engineering, 82*, 241–254.

Garofalo, F., Foti, S., Hollender, F., Bard, P. Y., Cornou, C., Cox, B. R., … Yamanaka, H. (2016a). InterPACIFIC project: comparison of invasive and non-invasive methods for seismic site characterization. Part I: Intra-comparison of surface wave methods. *Soil Dynamics and Earthquake Engineering, 82*, 222–240.

Garofalo, F., Sauvin, G., Socco, L. V., & Lecomte, I. (2015). Joint inversion of seismic and electric data applied to 2D media. *Geophysics, 80*, EN93–EN104.

Garotta, R. (2000). *Shear waves from acquisition to interpretation. Distinguished Instructor Series, 3*. Tulsa: SEG.

Gazoty, A., Fiandaca, G., Pedersen, J., Auken, E., & Christiansen, A. V. (2013). Data repeatability and acquisition techniques for time-domain spectral induced polarization. *Near Surface Geophysics, 11*, 391–406.

Gebregziabher, B. (2011). *Environmental and engineering geophysical studies for sinkhole problems using seismic reflection, refraction tomography, electrical resistivity imaging, and joint inversions* (Ph.D. thesis). LU Hannover http://edok01.tib.uni-hannover.de/edoks/e01dh11/646420569.pdf.

Ge, H., & Jackson, M. P. A. (1998). Physical modeling of structures formed by salt with-drawal: Implications for deformation caused by salt dissolution. *AAPG Bulletin, 82*, 228–250.

Geoscientists *Without* Borders. (2016). http://www.seg.org/web/foundation/programs/geoscientists-without-borders/projects/sweden.

Ghose, R. (2002). High-frequency shear wave reflections from shallow subsoil layers using a vibrator source: sweep cross-correlation versus deconvolution with ground force derivative. In *Society of Exploration Geophysicists 72nd Annual International Meeting, Expanded Abstracts*.

Ghose, R., Carvalho, J., & Loureiro, A. (2013). Signature of fault zone deformation in near-surface soil visible in shear wave seismic reflections. *Geophysical Research Letters, 40*. http://dx.doi.org/10.1002/grl.50241.

GISOS. (2009). *Synthese des Traveaux de Recherche*. Apres Mines. http://www.gisos.org.

Goyer, G. G., & Watson, R. (1963). The laser and its application to meteorology. *Bulletin of the American Meteorological Society, 44*, 564–575.

Grgich, B. (2013). *The little volcano with a tummy ache*, 36 pp., ISBN:10:1481771973.

Gruber, S. (2015). In *3rd International Workshop on Geoelectrical Monitoring. GELMON 2015, Vienna, 24–26 November 2015*. Vienna: Berichte der Geologischen Bundesanstalt 112. ISSN 1017-8880, 88 p.

Gruber, S., Jihyang, C., & Minsun, K. (2013). In *2nd International Workshop on Geoelectrical Monitoring. GELMON 2015, Vienna, 4–6 December 2013.* Vienna: Berichte der Geologischen Bundesanstalt 104. ISSN 1017-8880, 72 p.
Gutiérrez, R., Gibeaut, J. C., Smyth, R. C., Hepner, T. L., Andrews, J. R., Weed, C., … Mastin, M. (2001). Precise airborne LIDAR surveying for coastal research and geohazards applications. *International Archives of Photogrammetry and Remote Sensing, 34*(part 3/W4), 185–192.
Haeberli, W., & Whiteman, C. (2014). *Snow and ice-related hazards, risks, and disasters.* Elsevier Amsterdam/Waltham/Oxford, ISBN 9780123948496.
Hanafy, S. M. (2010). Locating of sinkhole extension by seismic refraction tomography and GPR methods. In *72nd EAGE Conference and Exhibition Incorporating SPE EUROPEC 2010.* http://dx.doi.org/10.3997/2214-4609.201401144.
Hansen, L., L'Heureux, J. S., Sauvin, G., Polom, U., Lecomte, I., Vanneste, M., et al. (2013). Effects of mass-wasting on the stratigraphic architecture of a fjord-valley fill: correlation of onshore, shear-wave seismic and marine seismic data at Trondheim, Norway. *Sedimentary Geology, 289,* 1–18.
Hare, J., Ferguson, J., Aiken, C., & Brady, J. (1999). The 4-D microgravity method for waterflood surveillance: a model study for the Prudhoe Bay reservoir, Alaska. *Geophysics, 64,* 78–87.
Harp, E. L., & Jibson, R. W. (1995). Seismic instrumentation of land- slides: building a better model of dynamic landslide behavior. *B. Seismol. Soc. Am., 85,* 93–99.
Hayashi, K. (2014). Detailed S-wave velocity model and site response of the Hayward, East San Francisco Bay Area. In *SEG Annual Meeting Expanded Abstracts, 1996–1999.* http://dx.doi.org/10.1190/segam2014-0385.1.
Heincke, B., Green, A. G., van der Kruk, J., & Horstmeyer, H. (2005). Acquisition and processing strategies for 3D georadar surveying a region characterized by rugged topography. *Geophysics, 70,* K53–K61.
Heincke, B., Gunther, T., Dalsegg, E., Ronning, J. S., Ganerød, G. V., & Elvebakk, H. (2010). Combined three-dimensional electric and seismic tomography study on the Aknes rockslide in western Norway. *Journal of Applied Geophysics, 70,* 292–306.
Heincke, B., Maurer, H., Green, A. G., Willenberg, H., Spillmann, T., & Burlini, L. (2006). Characterizing an unstable mountain slope using shallow 2- and 3-D seismic tomography. *Geophysics, 71,* B241–B256.
Hilbich, C., Hauck, C., Hoelzle, M., Scherler, M., Schudel, L., Völksch, I., … Mäusbacher, R. (2008). Monitoring mountain permafrost evolution using electrical resistivity tomography: a 7-year study of seasonal, annual, and long-term variations at Schilthorn, Swiss Alps. *Journal of Geophysical Research, 113,* F01S90.
Hinze, W. J., von Frese, R. R. B., & Saad, A. H. (2013). *Gravity and magnetic exploration: principles, practices, and applications.* Cambridge University Press, ISBN 9780521871013, 525 pp.
Historic Earthquakes. (1964). *Great Alaska earthquake, Alaska,* 1964 March 28 03:36 UTC (1964 March 27 05:36 p.m. local time), Magnitude 9.2 http://earthquake.usgs.gov/earthquakes/states/events/1964_03_28.php.
Holliger, K., & Bergmann, T. (2002). Numerical modeling of borehole georadar data. *Geophysics, 67,* 1249–1257.
Hönig, M., & Tezkan, B. (2007). 1D and 2D Cole-Cole-inversion of time-domain induced-polarization data. *Geophysical Prospecting, 55,* 117–133.
Hons, M. S., Stewart, R. R., Lawton, D. C., Bertram, M. B., & Hauer, G. (2008). Field data comparisons of MEMS accelerometers and analog geophones. *The Leading Edge, 27,* 896–903.

Hornbach, M. J., Braudy, N., Briggs, R., Cormier, M.-H., Davis, M., Diebold, J., … Templeton, J. (2010). High tsunami frequency as a result of combined strike-slip faulting and coastal landslides. *Nature Geoscience, 3*, 783–788.

Hornbach, M. J., Mann, P., Frohlich, C., Ellins, K., & Brown, L. (2011). Assessing geohazards near Kingston Jamaica: Initial results from chirp profiling. *The Leading Edge, 30*. http://dx.doi.org/10.1190/1.3575287.

Hornby, B. E., Bostick, F. X., Williams, B. A., Lewis, K. L., & Garossino, P. G. (2005). Field test of a permanent in-well fiber-optic seismic system. *Geophysics, 70*, E11–E19.

Hunter, J. A., & Motezedian, D. (2006). *Shear wave velocity measurements for soft soil earthquake response evaluation in the eastern Ottawa region*. Ontario, Canada: EEGS, SAGEEP (Expanded Abstract).

Hunter, J. A., Pullan, S. E., Burns, R. A., Gagne, R. M., & Good, R. S. (1984). Shallow seismic-reflection mapping of the overburden-bedrock interface with the engineering seismo graph—Some simple techniques. *Geophysics, 49*, 1381–1385.

Hyslop, C., & Stewart, R. (2015). Imaging lateral heterogeneity using reflected surface waves. *Geophysics, 80*(3), EN69–EN82.

Independent. (2014). Last time available on July 4, 2016 http://www.independent.co.uk/news/world/asia/japan-volcano-30-hikers-feared-dead-after-sudden-eruption-9760213.html.

Jacobs, R. (2008). Geoscientists without borders: Schlumberger leads the way. *The Leading Edge, 27*, 292–293. http://dx.doi.org/10.1190/tle27030292.1.

Jacobs, R. (2015). Geoscientists without borders update — It is good news for all of us. *The Leading Edge, 34*, 996. http://dx.doi.org/10.1190/tle34090996.1.

Jakka, R. S., Roy, N., & Wason, H. R. (2014). Implications of surface save data measurement uncertainty on seismic ground response analysis. *Soil Dynamics and Earthquake Engineering*, 61–62, 239–245.

Jentzsch, G., Punongbayan, R. S., Schreiber, U., Seeber, G., Völksen, C., & Weise, A. (2001). Mayon volcano, Philippines: change of monitoring strategy after microgravity and GPS measurements from 1992 to 1996. *Journal of Volcanology and Geothermal Research, 109*, 219–234.

Jeppson, T. N., Bradbury, K. K., & Evans, J. P. (2010). Geophysical properties within the San Andreas fault zone at the San Andreas fault observatory at depth and their relationships to rock properties and fault zone structure. *Journal of Geophysical research, 115*, B12423.

Johansson, S., & Dahlin, T. (1996). Seepage monitoring in an earth embankment dam by repeated resistivity measurements. *European Journal of Engineering and Environmental Geophysics, 1*, 229–247.

Jol, H. M. (2008). *Ground penetrating radar theory and applications*. Elsevier, ISBN 978-0-444-53348-7.

Jones, E. J. W. (1999). *Marine geophysics*. Wiley, ISBN 978-0-471-98694-2, 474 pp.

Jongmans, D., & Garambois, S. (2007). Geophysical investigation of landslides: a review. *Bulletin de la Société Géologique de France, 178*(2), 101–112.

Jousset, P., Reinsch, T., Henninges, J., Blanck, H., & Ryberg, T. (2016). Strain and ground-motion monitoring at magmatic areas: ultra-long and ultra-dense networks using fibre optic sensing systems. *Geophysical Research Abstracts, 18*, EGU2016–15707.

Juhlin, C., Dehghannejad, M., Lund, B., Malehmir, A., & Pratt, G. (2010). Reflection seismic imaging of the end-glacial Parvie Fault system, northern Sweden. *J. Appl. Geophys., 70*, 307–316.

Kääb, A. (2005). Remote sensing of mountain glaciers and permafrost creep. *Schriftenreihe Physische Geographie, 48*, 266 pp., ISBN:3855432449.

Kääb, A., Huggel, C., Fischer, L., Guex, S., Paul, F., Roer, I., … Weidmann, Y. (2005). Remote sensing of glacier- and permafrost-related hazards in high mountains: an overview. *Natural Hazards and Earth System Science, 5*, 527–554, 1684-9981/nhess/2005-5-527.

Kalscheuer, T., Bastani, M., Donohue, S., Persson, L., Pfaffhuber, A. A., Reiser, F., & Ren, Z. (2013). Delineation of a quick clay zone at Smørgrav, Norway, with electromagnetic methods under geotechnical constraints. *Journal of Applied Geophysics, 92*, 121–136.

Kalscheuer, T., Garcia, M., Meqbel, N., & Pedersen, L. B. (2010). Non-linear model error and resolution properties from two-dimensional single and joint inversions of direct current resistivity and radiomagnetotelluric data. *Geophysical Journal International, 182*, 1174–1188.

Kaufmann, G., & Romanov, D. (2008). Cave development in the Swabian Alb, south Germany, A numerical perspective. *Journal of Hydrology, 349*(3–4), 302–317.

Kaufmann, G., Romanov, D., & Hiller, T. (2010). Modelling three-dimensional karst aquifer evolution using different matrix-flow contributions. *Journal of Hydrology, 388*, 241–250.

Kaufmann, G., Romanov, D., & Nielbock, R. (2011). Cave detection using multi-geophysical methods, the UnicornCave, Harz mountains, Germany. *Geophysics, 76*(3), B71–B77.

Kearey, P., Brooks, M., & Hill, I. (2002). *An introduction to geophysical exploration* (3rd ed.). Wiley-Blackwell, ISBN 978-0-632-04929-5, 272 pp.

Keaton, J. R., Wartman, J., Anderson, S., Benoit, J., deLaChapelle, J., Gilbert, R., et al. (2014). *The 22 March 2014 oso landslide, Snohomish County, Washington*. Geotechnical Extreme Events Reconnaissance (GEER) report.

Kelly, K. R. (1983). Numerical study of love wave propagation. *Geophysics, 48*, 833–853.

Kent, J. D., & Dunaway, L. (2013). Real-time GPS network monitors bayou corne sinkhole event. *Eos, Transactions American Geophysical Union, 94*(43), 385–386. http://dx.doi.org/10.1002/2013EO430002.

Kitsunezaki, C., & Muraoka, A. (2015). Geothermal activity of Akita-Komagatake volcano after the 1970 eruption observed by repeated geophysical measurements at fixed stations. *Bulletin of the Volcanological Society of Japan, 3*, 365–380.

Kobayashi, T., Morishita, Y., & Yarai, H. (2015). Detailed crustal deformation and fault rupture of the 2015 Gorkha earthquake, Nepal, revealed from ScanSAR-based interferograms of ALOS-2. *Earth Planets Space, 67*, 201. http://dx.doi.org/10.1186/s40623-015-0359-z.

Komatina, S., Suto, K., & Urosevic, M. (2015). Assessment of flood damaged infrastructures in bosnia & herzegovina and Serbia: an SEG geoscientists without borders project. In *Proceedings of SEG Annual Meeting Technical Program Expanded Abstracts* (pp. 4968–4972). http://dx.doi.org/10.1190/segam2015-5903074.1. New Orleans. Permalink:.

Komatina, S., Urosevic, M., & Suto, K. (2014). Geophysics for natural disasters – Balkan cries for help! *Preview Issue, 170*, 18–19 (Australian Society of Exploration Geophysicists).

Konstantaki, L. A., Carpentier, S., Garofalo, F., Bergamo, P., & Socco, L. V. (2013). Determining hydrological and soil mechanical parameters from multichannel surface wave analysis across the Alpine fault at Inchbonnie, New Zealand. *Near Surface Geophysics, 11*, 435–448.

Kottmeier, C., Agnon, A., Al-Halbouni, D., Alpert, P., Corsmeier, U., Dahm, T., … Weber, M. (2016). New perspectives on interdisciplinary earth science at the Dead Sea: the DESERVE project. *Science of the Total Environment, 544*, 1045–1058. http://dx.doi.org/10.1016/j.scitotenv.2015.12.003. Online ISSN: 0048-9697.

Kramer, S. L. (1996). *Geotechnical earthquake engineering*. New Jersey: Prentice Hall, 653 pp.

Krawczyk, C. M. (2013a). Urban geohazards – integrated approaches in applied geophysics for monitoring of sinkhole and landslide processes. In *Proceedings of GeoMonitoring 2013* (pp. 105–110). Hannover/Germany: Leibniz Universität.

Krawczyk, C. M. (2013b). *Auslaugungen im Untergrund und Erdfallstrukturen. Broschüre 'Heimliche Prozesses im Untergrund', Stiftung Umwelt und Schadenvorsorge* (pp. 8–9). http://www.

stiftung-schadenvorsorge.de/web/export/sites/stuv/downloads/broschuere_symposium_2013.pdf.

Krawczyk, C. M., & Dahm, T. (2011). *Charakterisierung und Überwachung von Salz-bezogenen Erdfällen in urbanen Gebieten*. White Paper. http://www.liag-hannover.de.

Krawczyk, C. M., & Dahm, T. (2016). *SIMULTAN: Sinkhole instability - Integrated multi-scale monitoring and analysis. Planet Erde*. http://www.planeterde.de/Members/huenken/BMBFProjekte/simultan.

Krawczyk, C. M., Henk, A., Tanner, D. C., Trappe, H., Ziesch, J., Beilecke, T., … Kolditz, O. (2015). *Seismic and sub-seismic deformation prediction in the context of geological carbon trapping and storage. Advanced technologies in earth sciences* (pp. 97–113). Switzerland: Springer, ISBN 978-3-319-13929-6. http://dx.doi.org/10.1007/978-3-319-13930-2_5.

Krawczyk, C. M., Polom, U., & Beilecke, T. (2013). Shear-wave reflection seismics as valuable tool for near-surface urban applications. *The Leading Edge, Special Volume on Urban Geophysics, 32*(3), 256–263. http://dx.doi.org/10.1190/tle32030256.1.

Krawczyk, C. M., Polom, U., & Buness, H. (2015). *Geophysikalische Schlüsselparameter zur Überwachung von Erdfällen – Stand und Ziele der aktiven Seismik*. DGG-Kolloquium Sonderband Georisiken/Erdfälle I/2015, p. 19–30, ISSN 0947-1944. Deutsche Geophysikalische Gesellschaft (DGG).

Krawczyk, C. M., Polom, U., & Dahm, T. (2012). Imaging of urban sinkhole structures - combination of P-wave and shear-wave reflection seismic profiling in the metropolitan region of Hamburg. *Geophysical Research Abstracts, 14*, EGU2012–3164.

Krawczyk, C. M., Polom, U., Malehmir, A., & Bastani, M. (2013). Quick-clay landslides in Sweden – insights from shear-wave reflection seismics and geotechnical integration. In *Proceedings of the near surface geoscience 2013, Bochum/Germany*. http://dx.doi.org/10.3997/2214-4609.20131348, 5 pp.

Krawczyk, C. M., Polom, U., Trabs, S., & Dahm, T. (2012). Sinkholes in the city of Hamburg – new urban shear-wave reflection seismic system enables high-resolution imaging of subrosion structures. *Journal of Applied Geophysics, 78*, 133–143.

Krawczyk, C. M., Polom, U., Wiyono, W., & Pramono, B. (2015). Tailoring reflection seismic experiments to geothermal exploration targets in Indonesia – Wayang Windu geothermal field case study. In *World Geothermal Congress 2015, Paper 13034* (pp. 1–8).

Kristensen, L., Rivolta, C., Dehls, J., & Blikra, L. H. (2013). GB InSAR measurement at the Åknes rockslide, Norway. In *Proceedings of the International Conference Vajont 1963–2013. Thoughts and Analyses after 50 years since the Catastrophic Landslide, Padua, Italy, 8–10 October 2013*.

van der Kruk, J., Revil, A., & Slob, E. (2010). Introduction to hydrogeophysics – electric and electromagnetic methods. *Geophysics, 75*, WA1–WA2. http://dx.doi.org/10.1190/1.3483599.

Kühn, D., Ohrnberger, M., Vollmer, D., & Dahm, T. (2011). Imaging a shallow salt diapir using ambient seismic vibrations beneath the densely built-up city area of Hamburg, Northern Germany. *Journal of Seismology, 15*, 507–531.

L'Heureux, J.-S., Hansen, L., Longva, O., Emdal, A., & Grande, L. (2010). A multidisciplinary study of submarine slides at the Nidelva fjord delta, Mid-Norway—implications for hazard assessments. *Norwegian Journal of Geology, 90*, 1–20.

L'Heureux, J. S., Hansen, L., Sauvin, G., Polom, U., Lecomte, I., & Vanneste, M. (2012). High resolution sub-surface geo-characterization from shear-wave seismic reflection profiling: an example from the Trondheim harbour, mid Norway. *Danish Geotechnical Society Bulletin, 27*, 141–149.

Labo, J. (1986). In S. H. Mentemeier, & C. A. Cleneay (Eds.), *A practical introduction to borehole geophysics: An overview of wireline well logging principles for geophysicists*. Society of Exploration Geophysicists. http://dx.doi.org/10.1190/1.9781560802587.

Laigle, M., Becel, A., de Voogd, B., Hirn, A., Taymaz, T., Ozalaybey, S., et al. (2008). A first deep seismic survey in the Sea of Marmara: deep basins and whole crust architecture and evolution. *Earth and Planetary Science Letters, 270*, 168–179.

Lato, M., Hutchinson, J., Diederichs, M., Ball, D., & Harrap, R. (2009). Engineering monitoring of rockfall hazards along transportation corridors: using mobile terrestrial LiDAR. *Nature Hazards and Earth System Sciences, 9*, 935–946.

Lebert, F., Bernardie, S., & Mainsant, G. (2011). Hydroacoustic monitoring of a salt cavity, an analysis of precursory events of the collapse. *Natural Hazards and Earth System Sciences, 11*, 2663–2675. http://dx.doi.org/10.5194/nhess-11-2663-2011.

Leslar, M., Andelin, E., & Perry, G. (2009). Using Mobile lidar to survey railway infrastructure. In *Proceedings of the Innovative Lidar Solutions Conference 2009, Vaughan, Ontario, Canada. June 2009.*

Linde, N., Binley, A., Tryggvason, A., Pedersen, L. B., & Revil, A. (2006). Improved hydrogeophysical characterization using joint inversion of cross-hole electrical resistance and ground-penetrating radar traveltime data. *Water Resources Research, 42*, W12404. http://dx.doi.org/10.1029/2006WR005131.

Locat, J., & Sanfaçon, R. (2000). Multibeam surveys: a major tool for geosciences. In *Proceedings of the Canadian Hydraugraphics Conference, Montréal, 2000.* http://www.landslides.ggl.ulaval.ca/saguenay/publi/LocatJ.pdf.

Löfroth, H., & Kjellberg, U. (2003). *The May 1997 landslide in soft clay at Vagnhärad, Sweden, Lessons learnt from landslide disasters in Europe: European Commission Joint Research Centre* (pp. 33–40). Report EUR20558EN.

Löfroth, H., Suer, P., Dahlin, T., Leroux, V., & Schälin, D. (2011). *Quick-clay mapping by resistivity-surface resistivity, CPTU-R and chemistry to complement other geotechnical sounding and sampling: Swedish Geotechnical Institute, report GÄU 30.*

Loke, M. H. (2001). *Tutorial: 2-D and 3-D electrical imaging surveys.* Malaysia: Geotomo Software.

Loke, M. H., Chambers, J. E., Rucker, D. F., Kuras, O., & Wilkinson, P. B. (2013). Recent developments in the direct-current geoelectrical imaging method. *Journal of Applied Geophysics, 95*, 135–156.

Loke, M. H., Wilkinson, P. B., Uhlemann, S. S., Chambers, J. E., & Oxby, L. S. (2014). Computation of optimized arrays for 3-D electrical imaging surveys. *Geophysical Journal International, 199*, 1751–1764.

Long, M., Donohue, S., L'Heureux, J.-S., Solberg, I.-L., Rønning, J. S., Limacher, R., ... Lecomte, I. (2012). Relationship between electrical resistivity and basic geotechnical parameters for marine clays. *Canadian Geotechnical Journal, 49*(10), 1158–1168.

Looms, M. C., Binley, A., Jensen, K. H., Nielsen, L., & Hansen, T. M. (2008). Identifying unsaturated hydraulic parameters using an integrated data fusion approach on cross-borehole geophysical data. *Vadose Zone Journal, 7*, 238–248.

Luke, B., Taylor, W., Calderón-Macías, C., Jin, X., Murvosh, H., & Wagoner, J. (2008). Characterizing anomalous ground for engineering applications using surface-based seismic methods. *The Leading Edge, 29*, 1544–1549.

Lundberg, E., Malehmir, A., Juhlin, C., Bastani, M., & Andersson, M. (2014). High-resolution 3D reflection seismic investigation over a quick-clay landslide scar in southwest Sweden. *Geophysics, 79*, B97–B107.

Lundström, K., Larsson, R., & Dahlin, T. (2009). Mapping of quick clay formations using geotechnical and geophysical methods. *Landslides, 6*, 1–15.

Malehmir, A., & Bellefleur, G. (2009). 3D seismic reflection imaging of volcanic-hosted massive sulfide deposits: Insights from re-processing Halfmile Lake data, New Brunswick, Canada. *Geophysics, 74*, B209–B219.

Malehmir, A., Bellefleur, G., & Müller, C. (2010). 3D diffraction and mode-converted scattering signatures of base-metal deposits, Bathurst Mining Camp, Canada. *First Break, 28*, 41–45.

Malehmir, A., Dahlin, P., Lundberg, E., Juhlin, C., Sjöström, H., & Högdahl, K. (2011). Reflection seismic investigations in the Dannemora area, central Sweden: insights into the geometry of poly-phase deformation zones and magnetite-skarn deposits. *Journal of Geophysical Research, 116*, B11307.

Malehmir, A., Durrheim, R., Bellefleur, G., Urosevic, M., Juhlin, C., White, D., Milkereit, B., & Campbell, G. (2012). Seismic methods in mineral exploration and mine planning: a general overview of past and present case histories and a look into the future. *Geophysics, 77*, WC173–WC190.

Malehmir, A., Juhlin, C., Wijns, C., Urosevic, M., Valasti, P., & Koivisto, E. (2012). 3D reflection seismic investigation for open-pit mine planning and exploration in the Kevitsa Ni-Cu-PGE deposit, Northern Finland. *Geophysics, 77*, WC95–WC108.

Malehmir, A., Saleem, M. U., & Bastani, M. (2013). High-resolution reflection seismic investigations of quick-clay and associated formations at a landslide scar in southwest Sweden. *Journal of Applied Geophysics, 92*, 84–102.

Malehmir, A., Bastani, M., Krawczyk, C. M., Gurk, M., Ismail, N., Polom, U., & Persson, L. (2013). Geophysical assessment and geotechnical investigation of quick-clay landslides – a Swedish case study. *Near Surface Geophysics, 11*, 341–350.

Malehmir, A., Wang, S., Lamminen, J., Brodic, B., Bastani, M., Vaittinen, K., ... Place, J. (2015). Delineating structures controlling sandstone-hosted base-metal deposits using high-resolution multicomponent seismic and radio-magnetotelluric methods: a case study from Northern Sweden. *Geophysical Prospecting, 63*, 774–797.

Malehmir, A., Zhang, F., Dehgahnnejad, M., Lundberg, E., Döse, C., Friberg, O., ... Möller, H. (2015). Planning of urban underground infrastructure using a broadband seismic landstreamer—Tomography results and uncertainty quantifications from a case study in southwest of Sweden. *Geophysics, 80*, B177–B192.

Malehmir, A., Andersson, M., Mehta, S., Brodic, B., Munier, R., Place, J., ... Lund, B. (2016). Post-glacial reactivation of the Bollnäs fault, central Sweden - a multidisciplinary geophysical investigation. *Solid Earth, 7*, 509–527.

Malovichko, D., Dyagilev, R., Shulakov, D. Y., & Butyrin, P. (2009). Seismic monitoring of large-scale karst processes in a potash mine. In C. Tang (Ed.), *Controlling seismic hazard and sustainable development of deep mines* (Vol. 2, pp. 989–1002). New York: Rinton Press.

Massonnet, D., & Feigl, K. L. (1998). Radar interferometry and its application to changes in the Earth's Surface. *Reviews of Geophysics, 36*(4), 441–500.

Maurer, H., & Hauck, C. (2007). Instruments and methods – geophysical imaging of alpine rock glaciers. *Journal of Glaciology, 53*, 110–120.

Méric, O., Garambois, S., Jongmans, D., Vengeon, J. M., Wathelet, M., & Châtelain, J. L. (2005). Application of geophysical methods for the investigation of the large gravitational mass movement of Sechilienne (France). *Canadian Geotechnical Journal, 42*, 1105–1115.

Merz, K., Green, A. G., Buchli, T., Springman, S. M., & Maurer, H. (2015). A new 3D thin-skinned rock glacier model based on helicopter GPR results from the Swiss Alps. *Geophysical Research Letters*. http://dx.doi.org/10.1002/2015GL063951.

Merz, K., Maurer, H., Buchli, T., Horstmeyer, H., Green, A. G., & Springman, S. M. (2015). Evaluation of ground-based and helicopter ground-penetrating radar data acquired across an Alpine rock glacier. *Permafrost and Periglacial Processes*. http://dx.doi.org/10.1002/ppp.1836. n/a-n/a.

Merz, K., Maurer, H., Rabenstein, L., Buchli, T., Springman, S. M., & Zweifel, M. (2016). Multidisciplinary geophysical investigations over an alpine rock glacier. *Geophysics, 81*, WA147–WA157.

Mikko, H., Smith, C. A., Lund, B., Ask, M., & Munier, R. (2015). LiDAR-derived inventory of 25 post-glacial fault scarps in Sweden. *Journal of the Geological Society of Sweden, 137*, 334–338. http://dx.doi.org/10.1080/11035897.2015.1036360.

Miller, D. J. (1960). *Giant waves in Lituya Bay, Alaska. Geological survey professional papers.* University of Wisconsin Green Bay. USGS PP 354-C. Archived from the original on November 23, 2004. Available at: http://web.archive.org/web/20041123181612/http://www.uwsp.edu/geo/projects/geoweb/participants/dutch/LituyaBay/Lituya0.HTM.

Miller, R. D. (2003). High-resolution seismic-reflection investigation of a subsidence feature on U.S. Highway 50 near Hutchinson, Kansas. In K. S. Johnson, & J. T. Neal (Eds.), *Evaporite karst and engineering/environmental problems in the United States. Oklahoma geological survey circular 109* (pp. 157–167).

Miller, C. R., Routh, P. S., Brosten, T. R., & McNamara, J. P. (2008). Application of time-lapse ERT imaging to watershed characterization. Geophysics. *73*, G7–G17.

Miller, R. D., Ivanov, J., Xia, J., Peterie, S. L., & Sloan, S. L. (2010). Seismic investigations of subsidence hazards. *SEG Technical Program Expanded Abstracts, 2010*, 3769–3773. http://dx.doi.org/10.1190/1.3513634.

Miller, R. D., Villella, A., & Xia, J. (2008). Shallow high-resolution seismic reflection to delineate upper 400 m around a collapse feature in central Kansas. *Environmental Geosciences, 4*, 119–126.

Miller, R. D., Villella, A., Xia, J., & Steeples, D. W. (2005). Seismic investigation of a salt dissolution feature in Kansas. In D. K. Butler (Ed.), *Near-surface geophysics* (pp. 681–694). Society of Exploration Geophysicists. Investigations in Geophysics No. 13.

Miller, R. D., Xia, J., Harding, R. S., Neal, J. T., Fairborn, J. W., & Steeples, D. W. (1995). Seismic investigation of a surface collapse feature at Weeks Island Salt Dome, Louisiana. *AAPG Division of Environmental Geosciences Journal, 2*(2), 104–112.

Miller, R. D., Xia, J., & Park, C. B. (2005). Seismic techniques to delineate dissolution features (karst) at a proposed power plant site. In D. K. Butler (Ed.), *Near-surface geophysics* (pp. 663–679). Society of Exploration Geophysicists. Investigations in Geophysics No. 13.

Miller, R. D., Xia, J., Park, C. B., & Ivanov, J. M. (1999). Multichannel analysis of surfaces waves to map bedrock. *Leading Edge, 18*(12), 1392–1396.

Miller, R. D., Xia, J., & Steeples, D. W. (2009). Seismic reflection characteristics of naturally-induced subsidence affecting transportation. *Journal of Earth Science, 20*(3), 496–512.

Monserrat, O., Crosetto, M., & Luzi, G. (2014). A review of ground-based SAR interferometry for deformation measurement. *ISPRS Journal of Photogrammetry and Remote Sensing, 93*, 40–48. http://dx.doi.org/10.1016/j.isprsjprs.2014.04.001.

Moore, J. R., Gischig, V., Button, E., & Loew, S. (2010). Rockslide deformation monitoring with fiber optic strain sensors. *Nature Hazards and Earth System Sciences, 10*, 191–201.

Morey, D., & Schuster, G. T. (1999). Palaeoseismicity of the Oquirrh fault, Utah from shallow seismic tomography. *Geophysical Journal International, 138*, 25–35.

Nadim, F., Pedersen, S. A. S., Schmidt-Thomé, P., Sigmundsson, F., & Engdahl, M. (2008). Natural hazards in Nordic countries. *Episodes, 31*, 176–184.

Nagaoka, Y., Nishida, K., Aoki, Y., Takeo, M., & Ohminato, T. (2012). Seismic imaging of magma chamber beneath an active volcano. *Earth and Planetary Science Letters, 333-334*, 1–8.

Nasseri-Moghaddan, A., Cascante, G., & Hutchinson, J. (2005). A new quantitative procedure to determine the location and embedment depth of a void using surface waves. *Journal of Environmental and Engineering Geophysics, 10*, 51–64.

Naujoks, M., Weise, A., Kroner, C., & Jahr, T. (2008). Detection of small hydrological variations in gravity by repeated observations with relative gravimeters. *Journal of Geodesy, 82*, 543–553.

NIFS. (2015). *Detection of brittle materials. Summary report with recommendations.* Final report. NIFS Report no. 126/2015 www.naturfare.no.

Okada, H. (2003). *The microtremor survey method (translated by koya suto)*. Geophysical Monograph Series No. 12 (p. 135). Tulsa, OK, USA: Society of Exploration Geophysicists.
Olsson, P.-I., Dahlin, T., Fiandaca, G., & Auken, E. (2015). Measuring time domain spectral induced polarization in the on-time — decreasing the acquisition time and increasing the signal levels. *Journal of Applied Geophysics, 123*, 316—321.
Olsson, P.-I., Fiandaca, G., Dahlin, T., & Auken, E. (2015). Impact of time-domain IP pulse length on measured data and inverted models. In *Procs. Near Surface Geoscience 2015, 6—10 September 2015, Torino, Italy*, 5p.
Olsson, P.-I., Fiandaca, G., Juul Larsen, J., Dahlin, T., & Auken, E. (2016). Doubling the spectrum of time-domain induced polarization by harmonic de-noising, drift correction, spike removal, tapered gating and data uncertainty estimation. In *Procs. 4th International Workshop on Induced Polarization, 6—8 June 2016, Aarhus, Denmark*, 4p.
Paillet, F. L. (2001). Borehole geophysical applications in karst hydrogeology. In E. L. Kuniansky (Ed.), *U.S. Geological Survey Karst Interest Group Proceedings, Water-Resources Investigations Report 01-4011* (pp. 116—123).
Paine, J. G., Buckley, S., Collins, E. W., Wilson, C. R., & Kress, W. (2009). Assessing sinkhole potential at wink and daisetta using gravity and radar interferometry. In *Proceedings, 22nd Symposium on the Application of Geophysics to Engineering and Environmental Problems, Fort Worth, Texas, March 29—April 2* (pp. 480—488).
Parise, M., & Lollino, P. (2011). A preliminary analysis of failure mechanisms in karst and man-made underground caves in Southern Italy. *Geomorphology, 134*, 132—143.
Park, C. B., Miller, R. D., & Xia, J. (1999). Multichannel analysis of surface waves. *Geophysics, 64*, 800—808.
Park, J., Morgan, J. K., Zelt, C. A., & Okubo, P. G. (2009). Volcano-tectonic implications of 3-D velocity structures derived from joint active and passive source tomography of the island of Hawaii. *Journal of Geophysical Research, 114*, B09301. http://dx.doi.org/10.1029/2008JB005929.
Parolai, S., & Galiana-Merino, J. J. (2006). Effect of transient seismic noise on estimates of H/V spectral ratios. *Bulletin of the Seismological Society of America, 96*, 228—236. http://dx.doi.org/10.1785/0120050084.
Pedersen, L. B., Bastani, M., & Dynesius, L. (2006). Some characteristics of the electromagnetic field from radio transmitters in Europe. *Geophysics, 71*, G279—G284.
Pelton, J. R. (2005). Near-surface seismology. Surface-based methods. In D. K. Butler (Ed.), *Near-surface geophysics* (pp. 219—264). Tulsa, USA: Soc.. of Expl. Geophys.
Petley, D. N. (2012). Global patterns of loss of life from landslides. *Geology, 40*, 927—930. http://dx.doi.org/10.1130/G33217.1.
Pfaffhuber, A. A., & Bazin, S. (2011). Supporting geotechnical investigations with a 12-channel, time domain resistivity & induced polarization system; successes & pitfalls for settlement assessment, aquifer protection and rock quality mapping. In *DGG-Seminar "Ingenieurgeophysik", 6—8 April, Neustadt/W, Germany.*
Pfaffhuber, A., Bazin, S., & Helle, T. E. (2013). An integrated approach to quick-clay mapping based on resistivity measurements and geotechnical investigations. In L. Heureux, et al. (Eds.), *Advances in natural and technological hazards research, 36Landslides in sensitive clays: From geoscience to risk management*. Springer, 2013.
Pfaffhuber, A. A., Grimstad, E., Domaas, U., Auken, E., Halkjær, M., & Foged, N. (2010). Airborne EM mapping of rock slides and tunneling hazards. *The Leading Edge, 29*, 956—959.
Piatti, C., Foti, S., Socco, L. V., & Boiero, D. (2013). Building 3D shear velocity models using surface wave testing: the Tarcento Basin case history. *Bulletin of the Seismological Society of America, 103*(2A), 1038—1047. http://dx.doi.org/10.1785/0120120089.
Picozzi, M., Milkereit, C., Parolai, S., Jaeckel, K.-H., Veit, I., Fischer, J., & Zschau, J. (2010). GFZ Wireless Seismic Array (GFZ-WISE), a wireless mesh network of seismic sensors:

new perspectives for seismic noise array investigations and site monitoring. *Sensors, 10*, 3280–3304. http://dx.doi.org/10.3390/s100403280.
Place, J. A. P., & Malehmir, A. (2016). Using supervirtual first arrivals in controlled-source hardrock seismic imaging — well worth the effort. *Geophysical Journal International, 206*, 716–730. http://dx.doi.org/10.3997/2214-4609.201600030.
Place, J. A. P., Malehmir, A., Högdahl, K., Juhlin, C., & Persson Nilsson, K. (2015). Seismic characterization of the Grängesberg iron deposit and its mining-induced structures, central Sweden. *Interpretation, 3*, SY41–SY56.
Polom, U., Arsyad, I., & Kümpel, H.-J. (2008). Shallow shear-wave reflection seismics in the tsunami struck Krueng Aceh River Basin, Sumatra. *Advances in Geosciences, 14*, 135–140.
Polom, U., Bagge, M., Wadas, S., Winsemann, J., Brandes, C., & Krawczyk, C. M. (2013). Surveying near-surface depocentres by means of shear wave seismics. *First Break, 31*, 63–75. http://www.earthdoc.org.
Polom, U., Druivenga, G., Grossmann, E., Grüneberg, S., & Rode, W. (2011). *Transportabler Scherwellenvibrator: Deutsches Patent- und Markenamt.* Patentschrift DE10327757B4.
Polom, U., Hansen, L., Sauvin, G., L'Heureux, J.-S., Lecomte, I., Krawczyk, C. M., … Longva, O. (2010). High-resolution SH-wave seismic reflection for characterization of onshore ground conditions in the Trondheim harbor, central Norway. In R. D. Miller, J. D. Bradford, & K. Holliger (Eds.), *Advances in near-surface seismology and ground-penetrating radar, SEG, Tulsa* (pp. 297–312), ISBN 978-1-56080-224-2.
Polom, U., & Krawczyk, C. M. (2014). *Exploration of landslide geometry with seismics*. DGG-Kolloquium Sonderband Geohazards/Landslides I/2014, p. 41–50, ISSN 0947-1944. Deutsche Geophysikalische Gesellschaft (DGG).
Pringle, J. K., Styles, P., Howell, C. P., Branston, M. W., Furner, R., & Toon, S. M. (2012). Long-term time-lapse microgravity and geotechnical monitoring of relict salt mines, Marson, Cheshire, U.K. *Geophysics*, 77(6), B287–B294.
Pugin, A. J.-M., Brewer, K., Cartwright, T., Pullan, S. E., Didier, H. C., & Hunter, J. (2013). Near surface S-wave seismic reflection profiling-new approaches and insights. *First Break, 31*, 49–60.
Pugin, A. J. M., Larson, T. H., & Sargent, S. L. (2004). Near-surface mapping using SH-wave and P-wave seismic land-streamer data acquisition in Illinois, US. *The Leading Edge, 23*, 677–682.
Pugin, A. J.-M., Pullan, S. E., Hunter, J. A., & Oldenborger, G. A. (2009). Hydrogeological prospecting using P- and S-wave landstreamer seismic reflection methods. *Near Surface Geophysics*, 7, 315–327.
Rankka, K., Andersson-Sköld, Y., Hultén, C., Larsson, R., Leroux, V., & Dahlin, T. (2004). *Quick-clay in Sweden* (p. 137). Swedish Geotechnical Institute. Technical report number 65.
Re, S., Strobbia, C., De Stefano, M., & Virgilio, M. (2010). *Simultaneous joint inversion of refracted and surface waves: 80th SEG Conference and Exhibition. Denver, USA. Extended Abstracts.*
Reuther, C., Buurman, N., Kühn, D., Ohrnberger, M., Dahm, T., & Scherbaum, F. (2007). Erkundung des unterirdischen Raums der Metropolregion Hamburg: Das Projekt HADU (Hamburg-A Dynamic Underground). *Geotechnik, 30*, 11–20.
Reynolds, J. M. (2011). *An introduction to applied and environmental geophysics*. Wiley, ISBN 978-0-471-48536-0, 712 pp.
Roman, D. C., & Cashman, K. V. (2006). The origin of volcano-tectonic earthquake swarms. *Geology, 34*(6), 457–460.
Rømoen, M., Pfaffhuber, A., Karlsrud, K., & Helle, T. E. (2010). Resistivity on marine sediments retrieved from RCPTU —soundings: a Norwegian case study. In , *Vol. 2.*

International Symposium on Cone Penetration Testing, Huntington Beach CA USA, Proceedings (pp. 289–296).

Rosen, P., Hensley, S., Joughin, I., Li, F., Madsen, S., Rodriguez, E., & Goldstein, R. (2000). Synthetic aperture radar interferometry. In , *Vol. 88. Proceedings of the IEEE* (pp. 333–382).

Roth, D. L., Brodsky, E. E., Finnegan, N. J., Rickenmann, D., Turowski, J. M., & Badoux, A. (2016). Bed load sediment transport inferred from seismic signals near a river. *Journal of Geophysical Research Earth Surface, 121*. http://dx.doi.org/10.1002/2015JF003782.

Roy, N., SankarJakka, R., & Wason, H. R. (2013). Effect of surface wave inversion non-uniqueness on 1D seismic ground response analysis. *Natural Hazards, 68*, 1141–1153.

Rubin, Y., & Hubbard, S. S. (2006). *Hydrogeophysics* (Vol. 50). Springer Science & Business Media.

Rutishauser, A., Maurer, H. M., & Bauder, A. (2016). Helicopter-borne ground-penetrating radar investigations on temperate alpine glaciers: a comparison of different systems and their abilities for bedrock mapping. *Geophysics, 81*, WA119–WA129.

Rybakov, M., Goldshmidt, V., Fleischer, L., & Rotstein, Y. (2001). Cave detection and 4-D monitoring: a microgravity case history near the Dead Sea. *The Leading Edge, 20*, 896–900.

Rymer, H., Ferrucc, F., & Locke, C. A. (1998). Mount Etna: monitoring in the past, present and future. In D. J. Blundell, & A. C. Scott (Eds.), *Vol. 143. Lyell: The past is the key to the present* (pp. 335–347). Geological Society. London, Special Publications.

SafeLand. (2010). *Deliverable 4.1: Review of techniques for landslide detection, fast characterization, rapid mapping and long-term monitoring* (Edited for the SafeLand European project by Michoud, C., Abellán, A., Derron, M.-H., Jaboyedoff, M.).

SafeLand. (2011). *Deliverable 4.4: Guidelines for the selection of appropriate remote sensing technologies for monitoring different types of landslides* (Edited for the SafeLand European project by Stumpf, A., Kerle, N., Malet, J.-P.).

SafeLand. (2012). *Deliverable 4.8: Evaluation and development of reliable procedures and technologies for early warning* (Edited for the SafeLand European project by Bazin, S).

Salas-Romero, S., Malehmir, A., Snowball, I., Lougheed, B. C., & Hellqvist, M. (2016). Identifying landslide preconditions in Swedish quick clays – insights from integration of surface geophysical, core sample- and downhole property measurements. *Landslides, 13*(5), 905–923. http://dx.doi.org/10.1007/s10346-015-0633-y.

Sauvin, G., Lecomte, I., Bazin, S., Heureux, J., Vanneste, M., Solberg, I., & Dalsegg, E. (2013). Towards geophysical and geotechnical integration for quick-clay mapping in Norway. *Near Surface Geophysics, 11*, 613–623.

Schmidt, S., Wunderlich, J., Peters, A., & Heinke, O. (2013). Ingenieurgeologische Erkundung des Erdfalls vom 01. *Nov. 2010 am Rötbergrain in Schmalkalden und Beschreibung des Erdfallfrühwarnsystems in Schmalkalden (EIS 2010 bis 2012)*. Unveröff. Ergebnisbericht, Thür. Landesanst. f. Umwelt und Geologie, 179 S., 35 Anl., Weimar.

Seidel, K., & Serfling, U. (2010). Beitrag der Gravimetrie zur strukturellen Erkundung und zum Monitoring in Senkungsgebieten. In J. Gerardi (Ed.), *Vol. 244. Abgesoffene Bergwerke II, Staßfurt 2010. Erkennen, analysieren, bewerten und prognostizieren der zukünftigen Entwicklung der Bergbaufolgeschäden* (pp. 112–124). Hannover: Veröffentl. DGG.

Shan, C., Bastani, M., Malehmir, A., Persson, L., & Engdahl, M. (2014). Integrated 2D modeling and interpretation of geophysical and geotechnical data to delineate quick clays at a landslide site in southwest Sweden. *Geophysics, 79*, EN61–EN75.

Shan, C., Bastani, M., Malehmir, A., Persson, L., & Lundberg, E. (2016). Integration of controlled-source and radio magnetotellurics, electric resistivity tomography, and

reflection seismics to delineate 3D structures of a quick-clay landslide site in southwest Sweden. *Geophysics, 81*, B13–B29.
Sharma, P. V. (1997). *Environmental and engineering geophysics*. Cambridge University Press, ISBN 9780521576321, 475 pp.
Sheriff, R. E., & Geldart, L. P. (1995). *Exploration seismology*. Cambridge University Press, ISBN 9780521462822, 628 pp.
Siebert, L. (2004). Landslides resulting from structural failure of volcanoes, catastrophic landslides: effects, occurrence and mechanisms. In S. G. Evans, & J. V. DeGraff (Eds.), *Reviews in engineering geology* (Vol. XV). Boulder, Colorado: Geological Society of America.
Siripunvaraporn, W., & Egbert, G. (2000). An efficient data-subspace inversion method for 2D magnetotelluric data. *Geophysics, 65*, 791–803.
Sjödahl, P., Dahlin, T., & Johansson, S. (2009). Estimating seepage flow from resistivity monitoring data at the Sädva embankment dam. *Near Surface Geophysics*, 7, 463–474.
Sloan, S. D., Peterie, S. L., Ivanov, J., Miller, R. D., & McKenna, J. R. (2010). Void detection using near-surface seismic methods. In R. D. Miller, J. D. Bradford, & K. Holliger (Eds.), *Advances in near-surface Seismology and ground-penetrating radar* (pp. 201–218). Society of Exploration Geophysicists. Geophysical Developments Series No. 15.
Smith, C., Sundh, M., & Mikko, H. (2014). Surficial geologic evidence for early Holocene faulting and seismicity. *International Journal of Earth Sciences, 103*, 1711–1724.
Socco, L. V., Boiero, D., Foti, S., & Wisén, R. (2009). Laterally constrained inversion of ground roll from seismic reflection records. *Geophysics, 74*, G35–G45. http://dx.doi.org/10.1190/1.3223636.
Socco, L. V., Boiero, D., Foti, S., & Piatti, C. (2010). Advances in surface-wave and body-wave integration. In R. D. Miller, J. H. Bradford, & K. Holliger (Eds.), *SEG Geophysical Developments Series: Vol. 15. Advances in near-surface seismology and ground-penetrating radar* (pp. 55–73).
Socco, L. V., Foti, S., & Boiero, D. (2010). Surface-wave analysis for building near-surface velocity models — Established approaches and new perspectives. *Geophysics, 75*, 75A83–75A102.
Socco, L. V., Foti, S., Comina, C., & Boiero, D. (2012). Comment on "Shear wave profiles from surface wave inversion: the impact of uncertainty on seismic site response analysis. *Journal of Geophysics and Engineering, 9*, 241–243.
Socco, L. V., Jongmans, D., Boiero, D., Stocco, S., Maraschini, M., Tokeshi, K., & Hantz, D. (2010). Geophysical investigation of the sandalp rock avalanche deposits. *Journal of Applied Geophysics, 70*, 277–291.
Solberg, I. (2007). *Geological, geomorphological and geophysical investigations of areas prone to clay slides: Examples from Buvika, Mid Norway* (Ph.D. thesis). Norwegian University.
Solberg, I. L., Rønning, J. S., Dalsegg, E., Hansen, L., Rokoengen, K., & Sandven, R. (2008). Resistivity measurements as a tool for outlining quick-clay extents and valley-fill stratigraphy: a feasibility study from Buvika, Central Norway. *Canadian Geotechnical Journal, 45*, 210–225.
Sørensen, K. I., & Auken, E. (2004). SkyTEM – a new high-resolution helicopter transient electromagnetic system. *Exploration Geophysics, 35*, 191–199.
Supper, R., & Kauer, S. (2011). Geoelectric monitoring current research and perspectives for the future. In *International Workshop on Geoelectrical Monitoring. GELMON 2015, Vienna, 30 November – 2 December 2011*. Vienna: Berichte der Geologischen Bundesanstalt 92. ISSN 10178880, 277 p.
Supper, R., Ottowitz, D., Jochum, B., Kim, J.-H., Römer, A., Baron, I., et al. (2014). Geoelectrical monitoring: an innovative method to supplement landslide surveillance and early warning. *Near Surface Geophysics, 12*, 133–150.

Sutinen, R., Hyvönena, E., & Kukkonen, I. (2014). LiDAR detection of paleolandslides in the vicinity of the Suasselkä postglacial fault, Finnish Lapland. *International Journal of Applied Earth Observation and Geoinformation, 27*, 91–99.

Suto, K. (2016). An MASW survey for landslide risk assessment: a case study. *Valjevo in Serbia. Proceedings of 7th International Conference of Applied Geophysics, 82–86, Bangkok, Thailand.*

Suto, K., Arsenovic, S., Sugawara, J., & Aizawa, T. (2016). An investigation of a landslide-prone area by MASW and resistivity surveys. In *Proceedings of Symposium on the Application of Geophysics to Engineering & Environmental Problems (SAGEEP) 2016, Denver, USA.*

Suto, K., & Kristinof, R. (2014). An MASW survey to assess flood damaged road – a case history. In *Proceedings of Symposium on the application of Geophysics to engineering & environmental problems (SAGEEP) 2014, Boston, USA.*

Talebian, M., Fielding, E. J., Funning, G. J., Ghorashi, M., Jackson, J., Nazari, H., Parsons, B., Priestley, K., Rosen, P. A., Walker, R., & Wright, T. J. (2004). The 2003 Bam (Iran) earthquake: rupture of a blind strike-slip fault. *Geophysical Research Letters, 31*, L11611. http://dx.doi.org/10.1029/2004GL020058.

Tatsunori, I., Tsuji, T., & Matsuoka, T. (2013). Window-controlled CMP crosscorrelation analysis for surface waves in laterally heterogeneous media. *Geophysics, 78*, EN95–EN105. http://dx.doi.org/10.1190/GEO2013-0010.1.

Tertulliani, A., Leschiutta, I., Bordoni, P., & Milana, G. (2012). Damage distribution in L'Aquila city (Central Italy) during the 6 April 2009 earthquake. *Bulletin of the Seismological Society of America, 102*, 1543–1553.

The Arab Potash Company. (2000). *Forty-fourth Annual report*, 39 pp.; http://www.arabpotash.com.

The Florida Senate. (2010). *Interim Report 2011-104*, 47 pp. https://www.flsenate.gov

Tran, K. T., & Hiltunen, D. R. (2011). An assessment of surface wave techniques at the Texas A&M national geotechnical experimentation site. In C. H. Juang, et al. (Eds.), *Vol. 2011. GeoRisk (2011). Geotechnical risk assessment and management, 224. ASCEGSP* (pp. 859–866).

Tryggvason, A., Rögnvaldsson, S. T., & Flóvenz, Ò. G. (2002). Three-dimensional imaging of the P- and S-wave velocity structure and earthquake locations beneath Southwest Iceland. *Geophys. J. Int, 151*, 848–866.

Uhlemann, S., Wilkinson, P. B., Chambers, J. E., Maurer, H., Merritt, A. J., Gunn, D. A., & Meldrum, P. I. (2015). Interpolation of landslide movements to improve the accuracy of 4D geoelectrical monitoring. *Journal of Applied Geophysics, 121*, 93–105.

Urosevic, M., Komatina, S., Burazer, M., Suto, K., Arsenovic, S., Milosevic, D., ... Coren, F. (2016). *Geophysical characterization of landslides in Serbia and Bosnia and Herzegovina – A GWB project. Proceedings of 78th European Association of Geoscientists and Engineers.* Vienna: Austria.

USGS. (2013). About us http://landslides.usgs.gov/aboutus/.

USGS. (2014). *Earthquakes with 1,000 or more deaths 1900–2014.* http://earthquake.usgs.gov/earthquakes/world/world_deaths.php.

USGS. (2016a). *Lahar detection system. USGS: science for a changing world. U.S. geological survey.* http://volcanoes.usgs.gov/hazards/lahar/index.php.

USGS. (2016b). *The science of sinkholes.* http://www.usgs.gov/blogs/features/usgs_top_story/the-science-of-sinkholes.

Vanneste, M., Vardy, M. E., Knudsen, S., Mörz, T., Longva, O., Forsberg, C. F., ... Lunne, T. (2015). Integration of very-high-resolution seismics and CPTU data from a coastal area affected by shallow landsliding – the Finneidfjord natural Laboratory. In V. Meyer (Ed.), *Frontiers in Offshore Geotechnics III (ISFOG)* (pp. 1017–1022) (Oslo).

Viezzoli, A., Christiansen, A. V., Auken, E., & Sørensen, K. (2008). Quasi-3D modeling of airborne TEM data by spatially constrained inversion. *Geophysics, 65*(2), F105–F113.

Vöge, M., Frauenfelder, R., Ekseth, K., Arora, M. K., Bhattacharya, A., & Bhasin, R. K. (2015). The international archives of the photogrammetry, remote sensing and spatial information sciences, 2015. In *36th International Symposium on Remote Sensing of Environment, 11–15 May 2015, Berlin, Germany* (Vol. XL-7/W3).

Vosselman, G., & Maas, H.-G. (2012). *Airborne and terrestrial laser scanning*. Whittles Publishing, ISBN 978-1904445-87-6.

Wahlstrom, M., & Guha-Sapir, D. (2015). *The human cost of weather-related disasters 1995–2005*. http://www.unisdr.org/2015/docs/climatechange/COP21_WeatherDisastersReport_2015_FINAL.pdf.

Waltham, T., Bell, F., & Culshaw, M. (2005). *Sinkholes and subsidence*. Heidelberg: Springer, 382 pp.

Wang, S., Malehmir, A., & Bastani, M. (2016). Geophysical characterization of areas prone to quick-clay landslides in southwestern Sweden using radio-magnetotelluric and seismic methods. *Tectonophysics, 677–678*, 248–260.

Ward, S. H. (1990). Resistivity and induced polarization methods. In S. H. Ward (Ed.), *Geotechnical and environmental geophysics* (Vol. 1). Society of Exploration Geophysicists.

White, D. J., & Clowes, R. M. (2007). Shallow crustal structure beneath the Juan de Fuca Ridge from 2-D seismic refraction tomography. *Geophysical Journal International, 100*, 349–367.

Whitehead, K., & Hugenholtz, C. H. (2014). Remote sensing of the environment with small unmanned aircraft systems (UASs), part 1: a review of progress and challenges. *Journal of Unmanned Vehicle Systems, 2*(3), 69–85. http://dx.doi.org/10.1139/juvs-2014-0006.

Wikipedia (2015). https://en.wikipedia.org/wiki/2015_Shenzhen_landslide.

Williams, R. A., Simpson, R. W., Jachens, R. C., Stephenson, W. J., Odum, J. K., & Ponce, D. A. (2005). Seismic reflection evidence for a northeast-dipping Hayward fault near Fremont, California: Implications for seismic hazard. *Geophysical Research Letters, 32*, L13301.

Xia, J., Miller, R. D., Park, C. B., Hunter, J. A., & Harris, J. B. (2000). Comparing shear-wave velocity profiles from MASW with borehole measurements in unconsolidated sediments, Fraser river Delta, BC, Canada. *Journal of Environmental and Engineering Geophysics, 5*, 1–13.

Yilmaz, Ö. (1987). *Seismic data analysis: Processing, inversion, and interpretation of seismic data* (Vol. 2). Tulsa, USA: Society of Exploration Geophysicists.

Yilmaz, Ö. (2015). *Engineering seismology with applications to geotechnical engineering*. Society of Exploration Geophysicists, ISBN 978-1-56080-329-4, 964 pp.

Zerwer, A., Polak, M. A., & Santamarina, J. C. (2005). Detection of surface breaking cracks in concrete members using Rayleigh waves. *Journal of Environmental and Engineering Geophysics, 10*, 295–306.

Zhdanov, M. (2009). *Geophysical electromagnetic theory and methods*. Elsevier, ISBN 9780444529633, 868 pp.

Zollo, A., Gasparini, P., Virieux, J., Biella, G., Boschi, E., Capuano, P., … Mirabile, L. (1998). An image of Mt. Vesuvius obtained by 2D seismic tomography. *Journal of Volcanology and Geothermal Research, 82*, 161–173.

Zonge, K., Wynn, J., & Urquhart, S. (2005). Resistivity, induced polarization, and complex resistivity. In D. K. Butler (Ed.), *Near-surface geophysics – Investigations in geophysics* (pp. 265–300). Tulsa: SEG, ISBN 978-1-56080-130-6.

CHAPTER THREE

Light Snow Precipitation and Effects on Weather and Climate

I. Gultepe[*,§,1], R. Rabin[¶], R. Ware[||], M. Pavolonis[**]
[*]Environ. Canada, Toronto, Ontario, Canada
[§]NOAA NSSL, Norman, OK, United States
[¶]NOAA CIMSS, University of Wisconsin, Madison, WI, United States
[||]NCAR, CIRES, Boulder, CO, United States
[**]NOAA/NESDIS, Madison, WI, United States
[1]Corresponding author: E-mail: ismail.gultepe@ec.gc.ca

Contents

Advances in Geophysics, Volume 57
ISSN 0065-2687
http://dx.doi.org/10.1016/bs.agph.2016.09.001

Abstract

The objective of this work is to better understand light snow (LSN) precipitation measurements (precipitation rate (PR) < 0.5 mm/h) collected by optical present weather sensors (OPWS), weighing gauges, and spectral probes that are important for meteorological and hydrometeorological applications. Observations collected during the *Satellite Applications for Arctic Weather and Search and rescue (SAR) Operations* (SAAWSO) project that took place over Goose Bay, Newfoundland (NFL), Canada were studied to assess LSN characteristics and instrument sensitivities. Two case studies representing extreme environmental conditions temperature between 0 and −35°C, and snow occurrence for the SAAWSO project are presented. The ice crystal size and shape of LSN using a new platform called Ground Cloud Imaging Probe (GCIP) were obtained between 7.5 and 930 μm over 60 channels at 15 μm intervals. The measurements from the GCIP, Laser Precipitation Monitor (LPM), weighing gauges, and OPWS were used in the analysis. The results suggested the following: (1) LSN occurs at about 80% of time over the Arctic regions; (2) LSN can play a significant role in cooling at the surface and dehydration of the upper levels; and (3) OPWS can respond to LSN conditions better than weighing gauges. It is concluded that OPWS and spectral probes can improve measurement of LSN, including snow particle shape and size distribution with sizes <0.5 mm. Further research on LSN impact on weather and climate simulations is needed.

1. INTRODUCTION

Cloud particles have to be measured for size ranges less than a few 10s of μm in diameter to better verify the precipitation processes in operational applications and numerical models. An optimal approach to measure spectral characteristics of the cloud/fog particles and precipitation hydrometeors is to use optical sensors such as a GCIP (Ground Cloud Imaging Probe)[1]sensor, available from the Droplet Measurement Technologies, Inc. (DMT) (Gultepe et al., 2014b, 2014a, 2009), Optical Present Weather Sensors (OPWS) such as PWD52 from Vaisala Inc., Present Weather Sensor

[1] The nomenclature for acronyms is given in Appendix as Table A1.

(SWS) from Metek Inc., and a Laser Precipitation Monitor disdrometer (LPM), from Thies-Clima Inc. The latter two sensors have been used extensively in meteorological research (Brawn & Upton, 2008; Gultepe et al., 2014a, 2014b). The GCIP concept has been developed in Environment Canada (Gultepe et al., 2014a, 2014b, 2015) and designed by DMT Inc. Combination of these sensors covers the hydrometeor size range from 7.5 μm to a few cm, including light snow (LSN) precipitation particles (e.g., less than 500 μm). The LSN precipitation rate (PR_L) is defined as PR < 0.5 mm/h as these intensities are not measured accurately by weighing gauges because of a high-lower threshold value of PR as 0.1−0.5 mm/h and blowing wind conditions. Traditional optical sensors (e.g., disdrometers) are not capable of measuring LSN PR because of their weak optical response for sizes <200 μm (Brandes, Ikeda, Zhang, Schönhuber, & Rasmussen, 2007; Tapiador et al., 2012; Yang, Goodison, et al., 1999; Yang, Ishida, & Gunther, 1999).

The global distribution of snowfall is important for numerical weather prediction (NWP) and climate studies because of its effect on the hydrological cycle (Löhnert et al., 2011; Tapiador et al., 2012) and climate change. Low precipitation rates, low temperatures, and strong wind effects can make accurate snowfall measurements a challenge. Previous studies (Gultepe et al., 2015; Rasmussen, Vivekanandan, Cole, Myers, & Masters, 1999) suggested that the main challenges in adequately measuring snowfall are the high spatial and temporal variability as well as the enormous complexity of snow crystal habit, density, and particle size distribution (PSD). Accurate surface-based snowfall measurements are only sporadically available in the northern regions. Therefore, satellite remote sensing methods are needed to estimate LSN amount and rate but these methods lack sensitivity to low LSN PR. The operational stations usually provide total snowfall amount over the range of hours and they are subject to wind-induced errors (Sevruk & Klemm, 1989; Yang, Goodison, et al., 1999; Yang, Ishida, et al., 1999).

Over the Arctic regions, LSN precipitation cannot be adequately measured with traditional instruments, such as tipping buckets and weighing gauges, because the effects of wind, turbulence, and sensor sensitivity to SN measurements can be significant (Boudala, Rasmussen, Isaac, & Scott, 2014). Observational studies showed that the mean Arctic annual precipitation rate is about 1 mm/day (Randall et al., 2003; Roads, 2002), which is close to the uncertainty in weighing gauge point measurements (Goodison, Louie, & Yang, 1998; Gultepe, 2008). Although averaging scales are different, their results indicate the importance of daily LSN measurements. On average, trace recordings of snow can make up 45–50% of the annual total of precipitation days in the Arctic, with the monthly number of days reporting trace precipitation accounting for 15–80% of the monthly total number of precipitation days (Yang, Benson, & Ishida, 1998). Light precipitation amount (PA) can be very important for weather applications when heavy precipitation does not exist, and its characteristics can be calculated from the combination of the fog measuring device (FMD, with sizes up to 50 μm) and GCIP spectral measurements (Gultepe et al., 2015). In addition to instrument technical issues related to small particle size and mass detection, often extreme weather conditions such as low temperature, wind, and turbulence can also affect weighing gauges' measurement capabilities (WMO/CIMO, 1991).

Bogdanova, Ilyin, and Dragomilova (2002) analyzed Arctic missing precipitation based on Tretyakov shield measurements. They found that annual mean false precipitation comprises 30% more of the total measured precipitation. In their work they stated that blowing snow and blizzards significantly affect the in situ precipitation measurements (e.g., coastal high-latitudinal regions, ice sheets, tundra, mountain desert, and steppe climatic zones), accounting for the false precipitation and that is a much more difficult task to be accomplished. LSN PR is usually calculated by the measurements of OPWS [such as Vaisala present weather sensor (PWD) or Metek SWS] because of their lower threshold values for snow detection (0.01 mm/h) compared to the hotplate total precipitation sensor (TPS) and Geonor lower threshold of 0.5 mm/h without wind corrections (Rasmussen et al., 2012). Due to high winds and strong turbulence, LSN measurements over the Arctic regions TPS PR can also include large uncertainties (Boudala et al., 2014). For winds greater than 8 m/s, 1 mm/h threshold value is needed to obtain accurate PR for both the hotplate and Geonor 5-min averaged measurements (Rasmussen et al., 2012). The aforementioned studies suggest that LSN measurements need to be improved for better evaluation of precipitation forecast, and its understanding of the role of this precipitation on climate change.

Using Köppen's climate classification (Rohli & Vega, 2011) for the Polar Tundra (ET) and Polar Marine (EM) climates, Ballinger, Schmidlin, and Steinhoff (2013) showed that the EM has more precipitation than ET. According to these studies, the winter precipitation maximum over the EM is 90–100 mm/month for the Northern Hemisphere compared to the ET precipitation of <20 mm/month. This value is less than monthly average of 0.70 mm/day and it is comparable to LSN precipitation rate suggested by Gultepe et al. (2014b, 2014a).

Global re-analyses methods are often used for precipitation analysis. These methods use several fixed numerical weather prediction models and data assimilation schemes to produce gridded fields for PR over time periods suitable for climate research. Ballinger et al., (2013) conclude that caution must be exercised when using re-analysis data to study climate trends. Based on the re-analysis methods, climate change analysis is sensitive to the changes of the observing systems and processing methods (Bengtsson, Hagemann, & Hodges, 2004; Bengtsson, Hodges, & Hagemann, 2004; Sterl, 2004). Their work suggested that missing LSN amount can be important for annual mean precipitation assessments over cold climates when reanalysis techniques are used.

Gultepe et al. (2014b, 2014a) and Girard and Blanchet (2001a, 2001b) found that LSN usually occurs over the Arctic regions at temperatures less than $-10°C$. This is also emphasized by Intrieri and Shupe (2004). If LSN amount is not measured, underestimation of accumulated precipitation amount can occur. The LSN can be a high impact weather event as it can lower visibility down to a few hundreds of meters (Gultepe et al., 2015). In addition, the effect of LSN on the radiation budget needs to be accurately accounted for in climate models. A summary of global precipitation measurements including methods, uncertainty, datasets, and applications related to snow measurements, were studied by Gruber and Levizzani (2008) and Tapiador et al. (2012). These studies concluded that LSN measurements and its prediction may include large uncertainties, which can affect validation of model simulations with observations.

The falling ice fog particles which have size ranges from a few μm up to 300 μm (Gultepe et al., 2007, 2015) can also be considered as LSN. Their work suggest that visibility for aviation applications needs to be calculated for light snow and ice fog conditions because visibility versus LSN PR relationships can include large uncertainties when only snow ice water content (SIWC) is used (Stoelinga & Warner, 1999). Rasmussen et al. (1999) also suggested that snow type should be considered in the visibility versus snow PR calculations. For LSN conditions, visibility can go down to a few hundreds of meters for high ice crystal number concentration (N_i) with low SIWC (Gultepe et al., 2015). Their work suggested that visibility-PR relationships for SN conditions cannot be used accurately for aviation applications when ice crystal number concentration (N_i) is large. The above studies suggest that LSN properties should be measured accurately to improve visibility parameterizations.

In this work, observations collected during the *Satellite Applications for Arctic Weather and SAR (Search And Rescue) Operations* (SAAWSO) project that took place over the cold climatic regions of Canada, Goose Bay, Newfoundland, are used for accomplishing the following goals related to LSN precipitation and its effect on weather and climate issues:

- Evaluate LSN precipitation rate (PR) and amount (PA) for various optical and weighing sensors, and discuss an integrated particle spectra approach for LSN calculations;
- Compare PR and PA from optical probes with those of weighing gauges for two cases, and summarize the LSN occurrence results for a winter time period;
- Evaluate visibility versus PR relationships for LSN and SN conditions, and discuss their applicability in NWPs;
- Describe how newly derived light snow precipitation properties can be evaluated for LSN impact on the atmospheric heat budget.

2. INSTRUMENTS

The LSN relevant observations were collected by the ground-based in-situ instruments during the SAAWSO project. The project took place during the winter of 2012/2013 and 2013/2014 over one of the cold climatic regions of Canada, St. John's and Goose Bay, NFL, respectively, but only Goose Bay observations are analyzed due to usually wet snow conditions at the St. John's site. The instruments used in this work with their sensitivity for PR measurements are listed in Table 1 and they are summarized below.

2.1 Weighing Gauges

Weighing gauges (WGs) are designed based on weighing of falling precipitation received into a reservoir. When the gauge is operating below 0°C, an antifreeze blend was added to the bucket to melt frozen precipitation and to prevent freezing. Two types of antifreeze were usually used: Ethylene glycol and Propylene glycol. Two of them are used during the project for Geonor and Pluvio instruments, and these are described below.

2.1.1 Geonor Weighing Gauge

The precipitation collected in the Container (1500 mm capacity, collecting area of 200 cm^2) is weighed with three vibrating wire load sensors, which gives a frequency output (Tech. Manual, Geonor Inc. 2012). The frequency

Table 1 Shows precipitation and visibility measuring sensors used during satellite application for Arctic weather and SAR operations project took place in Goose Bay, NFL, Canada, during the winter of 2013–2014

Precipitation and Vis sensors	Manufacturing company	Measurements	Threshold PR&PA&Vis
GCIP	DMT	PS and shape	0.01 mm/h&−&> −50°C
PWD22	Vaisala	PT, PR	1.1 mm/min(0.05 mm/h)&
		PA	0.01 mm&10% &>−40°C&
		Vis	10 m (10%)
FD12p	Vaisala	PT, PR	1.2 mm/min(0.05 mm/h)&
		PA	0.01 mm&10% &>−40°C&
		Vis	10 m (10%)
SWS200	Biral	PT, PR, PA	0.0015 mm/ h&0.001 mm
		Vis	10 m, 5%&>−40°C
OSI-430	Optical Scientific	PT, PR, PA	0.01 mm/h&0.001 mm
		Vis	0.001 km&>−40°C
Sentry	Envirotech	Vis	10%, 30 m&>−40°C
LPM	Thiessen	PT, PS, PR, PA	0.005 mm/ h&0.005 mm >−40°C
		V_f	D[0.16->8 mm]; V_f[0.2 −20 m/s]
		Vis	Up to 30%
Geonor-200	Geonor	PR	0.05 mm/h&
		PA	0.05−0.1 mm&> −40°C
Pluvio	OTT	PR	12 mm/h or 0.20 mm/min&
		PA	0.10 mm; >−40°C

D, diameter; *PA*, amount; *PR*, precipitation rate; *PS*, particle spectra; *PT*, precipitation type; *Vis*, visibility; V_f, fall velocity.

will be a function of the applied tension on the wire, i.e., from this, the amount of precipitation can be computed. Using a three-sensor system ensures the continuation of data recording if one of the sensors should fail. This is made possible with a level-retaining device. It also ensures that the total amount of the precipitation will be measured even if there is a slight deviation from the horizontal plane. The frequency is recorded and the signal can easily be transmitted and interfaced to most data acquisition systems. Manufacturing accuracy is given as 0.1 mm/h but it can be as high as 0.5 mm/h. There are no moving mechanical parts, which eliminates other possible sources of error. Calculation of precipitation is described in the Analysis section.

2.1.2 Pluvio Weighing Gauge

The Pluvio precipitation gauge (OTT Co, Tech. Manual 01−0108) determines the weight of the collecting bucket including content every 6 s with a resolution of 0.01 mm (raw data). The difference between this measurement and the base weight of the empty collecting bucket gives the current bucket content. Every 6 s the OTT Pluvio2 calculates the bucket content using multiple raw values. A special filter algorithm prevents incorrect measurement resulting from effects such as wind. The difference between the current bucket content and the previous one gives the precipitation intensity in mm/min or mm/h. These 6-s values for the precipitation intensity are added to the accumulated precipitation amount. In contrast to conventional precipitation gauges, the OTT Pluvio2 works using the balance principle. The OTT Pluvio2 reliably recognizes precipitation, whether liquid or solid, by determining the weight of the collecting bucket. The OTT precipitation gauge is characterized in particular by its low maintenance needs. This is achieved by the OTT Pluvio2 with a high-capacity collecting bucket, the lack of a collection funnel or tilt mechanism and the extremely robust design of the balance mechanism. A high-precision, stainless steel load cell (hermetically sealed against environmental influences) that remains stable over a long period, is used as the sensor element. An integrated temperature sensor compensates for the temperature changes in the balance mechanism. The discrimination threshold for Pluvio2 with orifice opening of 200 cm^2 is 0.2 mm within an hour.

2.2 Optical Present Weather Sensors

Overall information on these sensors can be found in Table 1. The Optical Present Weather Sensors (OPWSs) work on the physical concept that extinction of the transmitted visible light by the hydrometeors is the function of hydrometeor optical properties and the decrease due to extinction is related to precipitation or fog spectral properties. These sensors usually are based on forward/backward scattering of transmitted visible light. Using a capacitor sensor called Vaisala DRD11A (FD12p Users Guide #M210296,

2002), snow precipitation rate and amount is calculated using an internal algorithm that is a function of temperature and the ratio of optical intensity to DRD11A intensity. In snow DRD11A intensity is proportional to water content of snow. The precipitation intensity calculated is based on both optical and capacitive measurements. However, due to undercatch, precipitation values because of wind induced errors might be lower than the real water content. The optical intensity is multiplied with a factor calculated from both DRD11A intensity and optical intensity to obtain snow water equivalent (SWE) amount.

The DRD11A operates via droplet detection rather than by a signal-level threshold. This sensor also features an analog rain signal for estimating rain intensity. As this signal is proportional to the percentage of moist or wet area on the sensor plate, rain intensity has a direct impact on the amplitude and variation of this analog signal. The DRD11A sensor is positioned at a 30 degrees angle. This design, together with the internal heating element, ensures that the surface dries quickly, an essential factor in calculating intensity. The same heating element also protects the surface from fog and condensed moisture, and is activated at low temperatures in order to melt snow, thus allowing snow detection.

The PWD22 (new version of FD12p), SWS200, and OSI-430 are built by companies such as Vaisala, Biral, and Optical Scientific, respectively. The Sentry built by Envirotech provides only visibility measurements and may not be used for precipitation measurements (Table 1). Note that SWS has the capability of measuring both forward and backward scattering of the transmitted light. The OSI-430 WIVIS visibility (Weather Identifier and visibility Sensor) optically measures precipitation induced scintillation and extinction coefficient and applies algorithms based on artificial intelligence and fuzzy logic techniques to automatically determine the precipitation type, precipitation rate, and visibility. The OSI visibility sensor measures precipitation by detecting the optical irregularities induced by particles falling through a beam of partially coherent infrared light (in the sample volume). These irregularities are known as scintillation. The twinkling of stars is a familiar example of scintillation. By detecting the intensity of the scintillations which are characteristic of precipitation, the precipitation rate is then determined. By analyzing the frequency spectrum of the induced scintillation, the precipitation type (i.e., rain, snow, etc.) is determined. All these sensors provide the NWS precipitation and hydrometeor codes that utilize the internal algorithms that may not be the same but they are all heated for icing and frost conditions.

2.3 Optical Spectral Precipitation Probes

The Optical Spectral Precipitation Probes (OSPPs) are designed based on the measurements of hydrometer physical characteristics. The Ground Cloud Imaging Probes (GCIPs) and Laser Precipitation Monitors (LPMs) used in this category are described below.

2.3.1 Ground Cloud Imaging Optical Probes

The Ground Cloud Imaging Probe (GCIP) concept for ground-based precipitations and ice fog measurements was developed in the Environment Canada and was based on the Droplet Measurement Technologies, Inc., Cloud Imaging Probe (CIP) designed for aircraft applications (Gultepe, 2008, pp. 59–82). The goal of GCIP use was to detect and measure light snow as well as ice and liquid fog microphysical parameters that can be used to support the measurements of disdrometers and fog devices. Fig. 1A and B shows GCIP and Fog Measuring Device (FMD), also called FS100 sensor, respectively, that have been used during the Fog and Remote Sensing Modeling (FRAM) and SAAWSO projects to study ice and freezing fog conditions (Gultepe et al., 2007, 2014a, 2015). The power distribution box part of aspirated GCIP is designed to be connected to a four wire 230 V AC with a 30 A circuit. The cloud probe instrument adapted from aircraft CIP probe operates from 28 V DC, 12 A, the blower from 115 V AC, 10 A, ¾ HP motor, and the inlet system contains three heaters 800 W each at 230 V AC. Each of the heated zones is temperature controlled via a temperature monitor and relay, with individual 15 breakers for protection. Two conductors are used for the power to the heater zones, and one conductor is allocated as a neutral/ground, which will bond to the inlet power distribution box. The blower operates from 115 V AC and has a 20 A breaker for control and protection. The blower has a standard line cord plug.

The GCIP uses an incident laser beam on hydrometeors ranging in size from 15 up to 930 μm which are optically magnified by a factor of 13.3, onto a 200 μm pitch, 64-element photo diode array (DMT Inc., 2004). A diode laser produces an oval 50 mW beam illuminating the diode array. Whenever a particle passes through the laser beam, a shadow is generated and imaged onto the diode array. On-board digital electronics begins storing photo-diode information at a clock rate associated with true air speed (TAS) frequency. Each sample of the 64 elements, taken at the TAS frequency, is called a slice up to 62 slices compose a particle image. Each particle detected

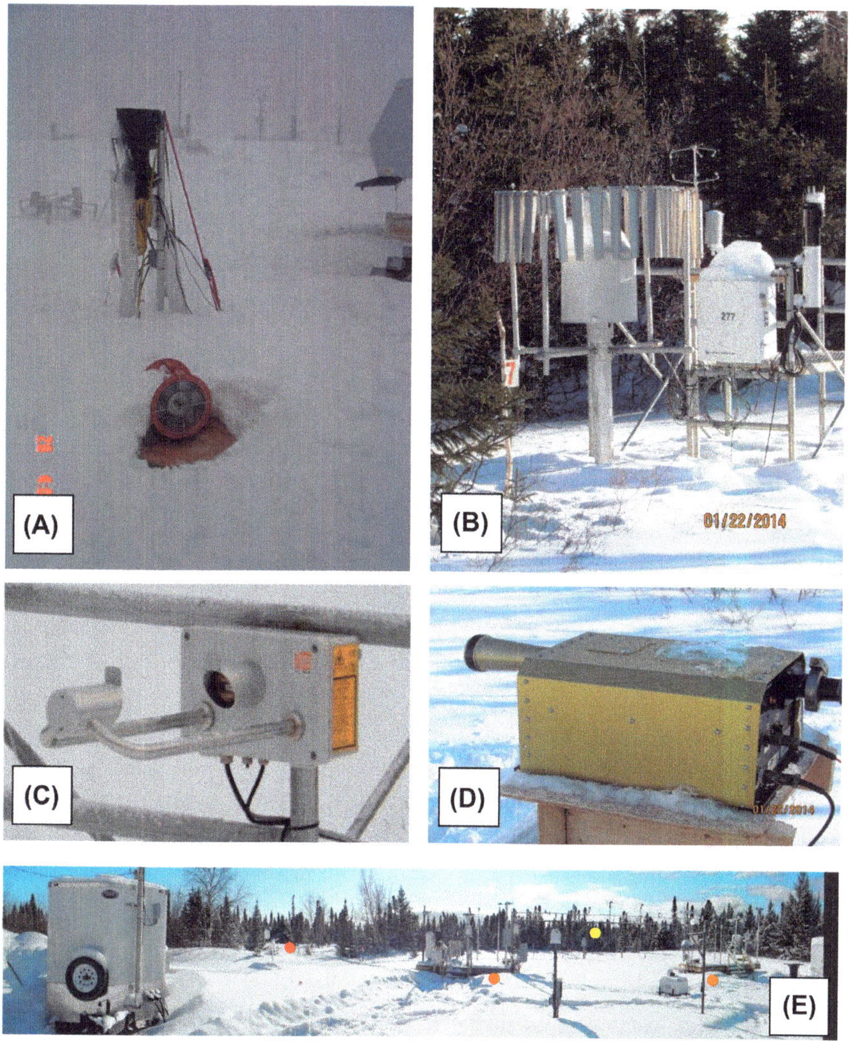

Figure 1 Box (A) shows details regarding the mounting of the ground cloud imaging probe. The airborne probe was fitted with an inverted bell shaped inlet which funneled the particles at 25 m/s through the measurement area of the imaging probe. The inlet was heated to prevent build-up of precipitation and icing. This picture is taken during satellite application for Arctic weather and SAR operations project took place in St. John's, NFL. Box (B) shows the Pluvio instrument with a single alter shield at 3 m surrounded with low-height trees and bushes. This site was 15 m away from the main site and it was used as reference snow measurements. Box (C) shows the laser precipitation monitor used for measuring snow particles with size range over 22 diameter class from 0.125 mm up to >8.0 mm and in 20 fall speed class from 0 up to >10 m/s. Box (D) shows the fog measuring device to measure fog particle spectra between 1 and 50 μm over 16 channels. Box (E) shows approximate locations of reference site (*red filled circles*), optical probes (*brown filled circles*), and unshielded Geonor (*yellow filled circle*).

has three shadowing densities at 25%, 50%, and 75% and shown as different colors. Particles shadowing an end diode (number 1 or 64) are rejected. The TAS is determined using an on-board pitot tube, providing air speed information at the instrument itself. The sampling area is 10 cm × 1.55 mm.

2.3.2 Laser Precipitation Monitor Disdrometer

The most common disdrometers used for snow precipitation measurements are Thies LPM and OTT Disdrometers (Gultepe et al., 2014a; Jaffrain & Berne, 2011). The LPM sensor used during this project is shown in Fig. 1C. The LPM (LPM Technical Manual, 2007) uses a laser-optical beaming source (laser diode and optics) that produces a parallel infrared (IR) light-beam (0.780 μm with 0.5 mW optical power, 40–47 cm^2 measuring area with x = 228 mm, y = 20 mm, and z = 75 mm. Its weight is 4.8 kg). A photo diode with a lens is situated on the receiver side in order to measure the optical intensity by transforming it into an electrical signal. When a precipitation particle falls through the light beam (measuring 45.6 cm^2 area), the receiving signal is reduced. The diameter of the particle is calculated from the amplitude. Moreover, the fall speed of the particle is determined from the duration of the reduced signal. The measured values are processed by a digital signal processor (DSP), and checked for plausibility (e.g., edge hits). Calculation comprises the intensity, quantity, type of precipitation (drizzle, rain, snow, hail as well as mixed precipitation), and the particle spectrum. Data is provided in 22 diameter class from 0.125 up to >8.0 mm and in 20 fall speed class from 0 up to >10 m/s.

2.4 Profiling Microwave Radiometer

A Profiling Microwave Radiometer (PMWR) provided continuous temperature and humidity profiles, and unique liquid water content profiles and these were used in assessing the clouds generating LSN conditions. The Radiometrics Corporation PMWR radiometer includes 5 K-band (22–30 GHz) and 7 V-band (51–59 GHz) microwave channels, a downward-looking external mount infrared μm(9.6–10.4 μm) radiometer and front-surface gold mirror assembly for cloud base temperature estimation, and surface temperature, relative humidity and pressure sensors. The Radiometrics 12-channel (model MP-3000) PMWR and its performance are described by Solheim, Godwin, and Ware (1998), Güldner and Spänkuch (2001), and in the WMO Guide (2010). The Radiometrics model MP-3000A, introduced in 2006, includes 35 microwave channels and an internal mount infrared sensor, providing improved accuracy and reliability (Cimini et al., 2011; Cimini, Nelson, Güldner, & Ware, 2015; Sanchez, Posada, Garcia-Ortega, Lopez, & Marcos, 2013; Ware et al., 2013).

Background error covariance analysis shows that Radiometrics PMWR models provide better temperature and humidity profile accuracy than numerical weather models to approximately 1 km and 3 km height, respectively (Cimini et al., 2011, 2015; Cimini, Westwater, & Gasiewski, 2010), with models showing higher accuracy at higher levels. When properly calibrated with appropriately trained neural networks, PMWR's obtain radiosonde-equivalent observation accuracy to 10 km height (Cimini et al., 2011; Güldner and Spänkuch, 2001; Knupp et al., 2009; Sanchez et al., 2013; Ware et al., 2013). The PMWR also provides 15% (Serke et al., 2014) to 50% (Ware et al., 2013) agreement with limited independent liquid profile and integrated liquid measurements and estimates (Politovich, Stankov, & Martner, 1995; Turner, 2007; Westwater, 1978).

3. MEASUREMENTS AND ANALYSIS

In this work, two cases representing light snow conditions that occurred during January 23 and May 3, 2014 are studied and the results are summarized for the entire data set to show the LSN occurrence compared to all snow cases. The LSN event on January 23 occurred with temperature (T) less than about −18°C. The LSN case on May 3 occurred with T close to 0°C. These cases were selected randomly to represent extreme weather conditions based on T with relatively low mean wind conditions (<2 m/s with gusts up to 4 m/s). If we had selected the higher wind conditions, our results would likely show severe effects on LSN measurements. Considering that blowing snow conditions usually happen when Uh > 7 m/s (Trouvilliez, Naaim-Bouvet, Bellot, Genthon, & Gallée, 2015), it is suggested that the impact of blowing snow effects on LSN measurements during project can be considered minimal.

A reference measurement sensor, Pluvio, for LSN measurements was set up using an Alter Shield (1937) within a bush-low tree environment. The Alter shield is a fence located around the inlet of the weighing gauge that reduces wind impact on precipitation measurements. Instrument measurement issues can be related to strong wind effects and low collection efficiencies, blowing snow and re-circulated precipitation, and optical responses. The snow precipitation measurement and its wind correction can easily be performed for SN rate >1 mm/h, but the same correction algorithm cannot work for LSN. In fact, calibration of optical and mechanical responses of the sensors for LSN particles were likely not performed over the very cold and dry environmental conditions, and, therefore, uncertainties in the cold environments can be more severe compared to these of mid-latitude environments. Zhang et al. (2015) compared the measurements of Geonor weighing gauge to those of the present weather sensors, and suggested that Geonor did not measure any snow when PA was less than 1 mm and Uh > 3.5 m/s. Even for Uh < 3.5 m/s, Zhang et al. suggested that Geonor recorded only about 44% of the total precipitation. During SAAWSO project, when PR < 0.5 mm/h, Geonor measurements did not detect the majority of the total snow precipitation. Similar to this finding, Rasmussen et al. (2012) had earlier suggested that the response of Geonor to LSN conditions was weak because its measurement sensitivity is not less than 0.5 mm/h under the stronger wind conditions.

3.1 Light Snow Calculation Using Hydrometeor Spectral Measurements

In the present work, hydrometeor spectral measurements were obtained using GCIP and LPM disdrometer. Total LSN PR (PR_T) based on GCIP and LPM measurements with sizes up to about 900 μm and cm size ranges, respectively, is obtained as

$$PR_T = A_c\left(\sum_{15\ \mu m}^{900\ \mu m} N(D)D^3\rho_s V_f(D)\Delta D + \sum_{900\ \mu m}^{cm} N(D)D^3\rho_s V_f(D)\Delta D\right), \tag{1}$$

where A_c is a constant (of value $3.6 \times 10^{+4}$) and V_f is the particle fall velocity which depends on particle density and shape (Heymsfield & Platt, 1984; Heymsfield et al., 2007; Kajikawa & Heymsfield, 1989). The ρ_s is the ice crystal density, V_f particle fall velocity, D the particle diameter, and N(D) the number density. The fall velocity and particle density are obtained from the GCIP measurements using empirical relationships. Light precipitation amount can be very important for weather applications where heavy precipitation is uncommon, and its characteristics can be calculated from the combination of the fog measuring device (FMD with sizes up to 50 μm) and GCIP spectral measurements (Gultepe et al., 2015). In addition to instrument technical issues related to small particle size and mass detection, extreme weather conditions such as low temperature, wind, and turbulence can also affect instrument measurement capabilities (WMO/CIMO, 1991).

The type of precipitation from LPM is determined from the statistics of all the particles as a function of both diameter and fall velocity. Based on manufacturer set up, a temperature sensor was included for discrimination of liquid particles when the temperature is >9°C and ice crystals when it is <−4°C. Between these temperatures, all forms of precipitation are possible.

3.2 Weighing Gauge Measurements and Uncertainties

Under the relatively calm wind conditions such as Uh < 5 m/s Geonor and Pluvio LSN measurements may not need wind corrections, but their sensitivity for LSN (<0.5 mm/h) can be an important issue. As described above, Pluvio with an Alter Shield in a bush-low tree environment (Fig. 1D) is used as reference but its PA measurements were usually either larger (or lower) than these of all optical probes for January 23 and May 3, 2014 cases. Both instruments are explained below. Geonor observations were not used with a shield, and its PR measurement uncertainity based on factory specification is given as 0.1 mm/h. This value can be up to 0.5 mm/h under turbulent and stronger wind conditions, or when ice crystals are small. The Geonor weighing gauge utilizes a three-vibrating wire technology (as described above) to distribute the snow mass equally. The following formula is used to obtain precipitation amount

$$\mathrm{PA} = \mathrm{A}(\mathrm{f} - \mathrm{f}_0) + \mathrm{B}(\mathrm{f} - \mathrm{f}_0)^2 \tag{2}$$

where PA is the accumulated precipitation (in cm), f is the frequency reading (in Hz), A and B calibration constants, and f_0 is the frequency with empty bucket (in Hz). Then, a 5-min average for the past data is applied to get PA at the current point. Noise from the data is removed by 10-min running averages. The sensitivity given by the manufacturer is 0.05 (0.1) mm for 600 (1000) mm.

Bias corrections of snow measurements for weighing gauges can be related to wind-induced undercatch, wetting loss, and evaporation loss (Goodison, 1981; Goodison et al., 1998; Sevruk & Klemm, 1989). Light snow conditions can be affected by all these biases but undercatch bias can be much stronger among these because of wind effects (Zhang, Ohata, Yang, & Davaa, 2004). They found that during the cold seasons, bias with either wetting or evaporation can be about 15% and with undercatch, it can be up to 20%.

In winter, evaporation and wetting losses together can be 0.10—0.20 mm/day (Aaltonen, Elomaa, Tuominen, & Valkovuori, 1993) and 0.15 mm/day (Sevruk, 1982), respectively, but these are not considered in the present work because mineral oil and antifreeze were used during entire project. Adding all of these biases may work out to about 0.5 mm/day.

The Pluvio-based PA measurements were used in obtaining LSN PR during which 60-min running averages were used to remove low response noises. Comparisons of PA from all sensors, including optical ones, are summarized in Section 5.

4. RESULTS

In this section, the results of GCIP, LPM, OPWS, and weighing instruments for LSN conditions are given and their importance with respect to the project goals are emphasized for two cases (Section 2), and then for the entire data set. The list of sensors and their sensitivity for LSN measurements are summarized in Table 1.

4.1 January 23, 2014 Light Snow Event

On this day, LSN occurred almost for the entire day and the total PA was about 1 mm and maximum PR based on LPM was less than 0.5 mm/h. To understand LSN formation, the profiling MWR (PMWR) thermodynamic and cloud observations are first summarized. During this LSN event, mean Uh was usually less than about 2 m/s

4.1.1 Profiling Microwave Radiometer Observations

A time-height cross section of Profiling Microwave Radiometer (PMWR) measurements can be used to analyze the LSN formation dynamics. Fig. 2 shows the PMWR time-height cross section to 1.2 km height of temperature (box A), relative humidity with respect to ice (RH_i, box B), and liquid water content (box C). Also shown are integrated water vapor (IWV) and integrated liquid water (ILW) retrievals (box D), radiometer surface temperature (T_{amb}), and infrared cloud base temperature (T_{ir}) measurements (box E).

Lower boundary layer diurnal variability as large as 6°C is seen in the temperature cross-section, peaking at 18:00 UTC (coordinated universal time) (15:00 local time). An inversion dominates the RH_i cross-section, peaking near 500 m height. Liquid water content (LWC) near 700 m height gradually increases starting at 06:00 UTC, exceeds 0.25 g/m^3 just

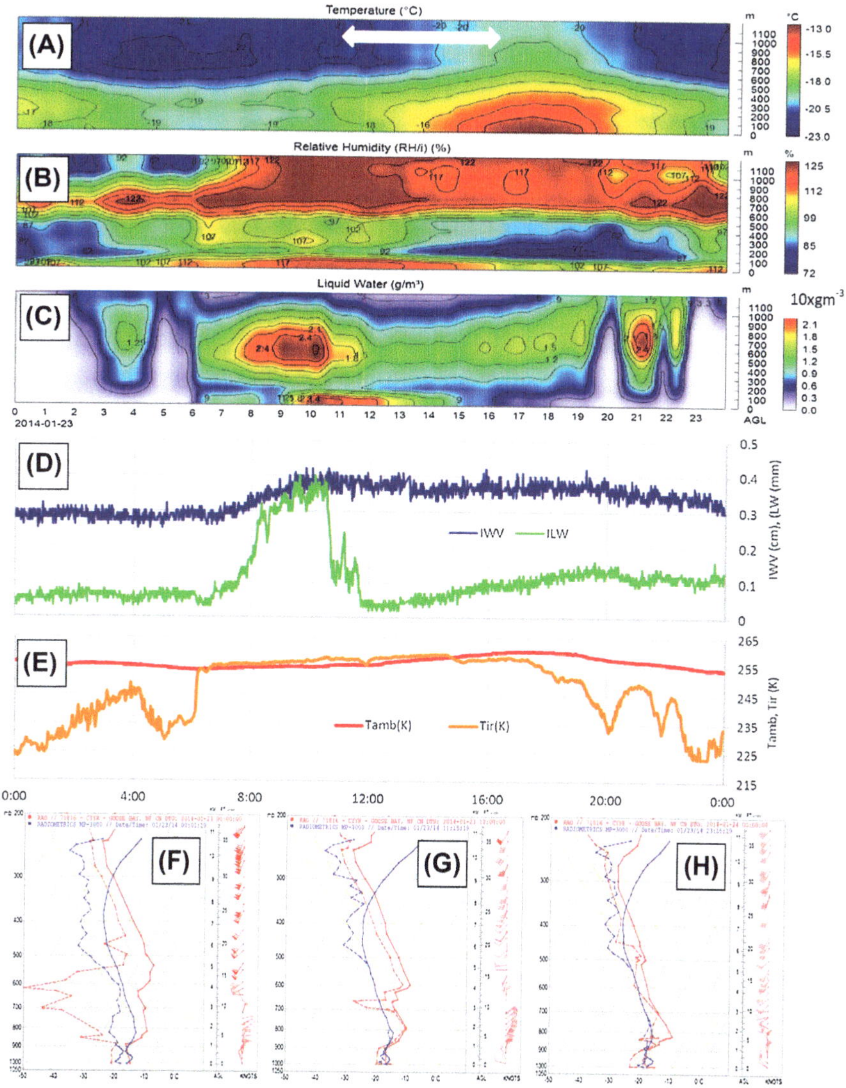

Figure 2 Radiometrics profiling microwave radiometer (PMWR) temperature (A), relative humidity over ice (B), and liquid water content (×10) profile retrievals to 1.2 km height (C), integrated water vapor (IWV) and integrated liquid water (ILW) retrievals (D), surface temperature (Tamb) and cloud base infrared temperature (Tir) (E), and Goose Bay radiosondes and PMWR profiles at 00Z, 12Z, 24Z are shown in (F—H), respectively, for January 23, 2014.

after 10:00 UTC, rapidly decreases to <1.5 g/m^3 by 12:00 UTC. It remains relatively constant until 19:00 UTC when it drops to near zero, and then, it grows to >0.25 g/m^3 near 21:00 UTC and drops to zero by 23:00 UTC. Indication of LWC existence by PMWR suggests that some liquid droplets were present during LSN formation.

IWV increases from 0.3 to 0.4 g/m^3 from 07:00 to 10:00 UTC (box D), correlated with ILW increase from 0.1 to nearly 0.4 g/m^3. Gradual ILW increase from 12:00 to 20:00 UTC and the lack of an increase near 21:00 UTC is inconsistent with the liquid density retrieval. We are unable to explain this inconsistency. T_{amb} and T_{ir} became closer in box E ($T_{amb} \sim T_{ir}$) from 06:00 to 17:00 UTC, suggesting of possible LSN conditions, but broken cloud cover is seen at other times.

Radiosonde and PMWR T and Td soundings up to 1 and 3 km height, respectively, show good agreement (boxes E, F, and G in Fig. 2), in particular when the high variability of radiosonde humidity point measurements is considered. Note that radiosonde data at higher levels, because of drifting with horizontal winds, do not represent soundings at the project site compared to MWR profiles. Upper level PMWR retrievals show a systematic bias consistent with Neural Network (NN) files trained with distant radiosondes (reiterating, Yarmouth is >1000 km SSW of Goose Bay). In addition, upper level retrieval accuracy is sensitive to PMWR calibration accuracy. Overall, LSN conditions were in good agreement with thermodynamic profiles.

4.1.2 In Situ Observations

Time series of precipitation rate for light snow obtained from the various precipitation sensors described in Table 1 are shown in Fig. 3. Although most of them are described as sensitive to PR down to 0.1 mm/h, this was not the case for many days having cold light snow measurements. The worst response to light snow conditions was from both Geonor and Pluvio (Fig. 3A) that were based on precipitation weighing methods. This was likely due to measurement limitations; sensitivity due to higher lower-threshold values and relatively stronger winds. The optical sensors were responding to LSN amount better than weighing gauges because of their lower threshold values (<0.01 mm/h) for particle detection and better response to small ice crystals. The Geonor, Pluvio, and FD12p PR usually were larger compared to the other measurements. The Geonor PR was randomly fluctuating; therefore, its measurements are neglected before 10 a.m. for PA calculations to be shown later. Both LPM and SWS

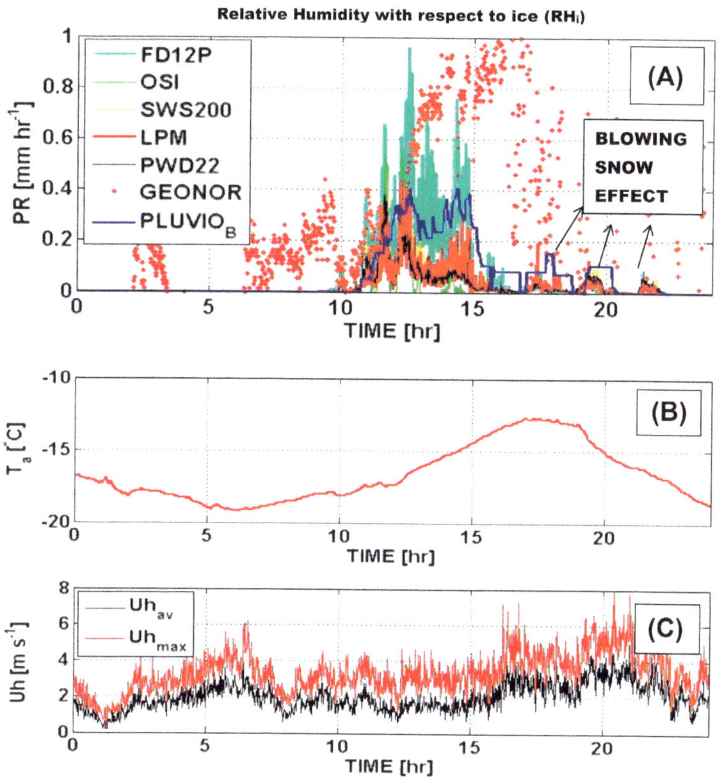

Figure 3 The comparison of light snow precipitation rate from various instruments (see legend, also in Table 1) for January 23, 2014 at T ~ −18°C occurred over Goose Bay. Blowing snow effects seen after 16:00 UTC (coordinated universal time) are consistent with Uh ~ 6 m/s are shown in box (A). T and Uh observations from WXT sensor over 1-min intervals (averaged and maximum) are shown in box (B and C), respectively.

measurements were usually in agreement with PWD; these instruments were measuring blowing snow crystals after 15:00 UTC (Fig. 3A) where Uh was up to 7–8 m/s (Fig. 3C). T values were between −15 and −20°C during the snow event (Fig. 3B). Clearly the optical sensors were responsive to small particles better than others. This was also shown by Gultepe et al. (2014b, 2014a, 2009) in which a fog device and FD12p sensor were used for ice fog detection. The Geonor was likely responding to wind conditions and reported the highest PR during LSN event. The Pluvio PR values were usually larger than measurements of the optical sensors, excluding FD12p.

Precipitation rate obtained from the subset of sensors given in Table 1 is compared to the PWD22 sensor (Fig. 4). This figure suggests that maximum

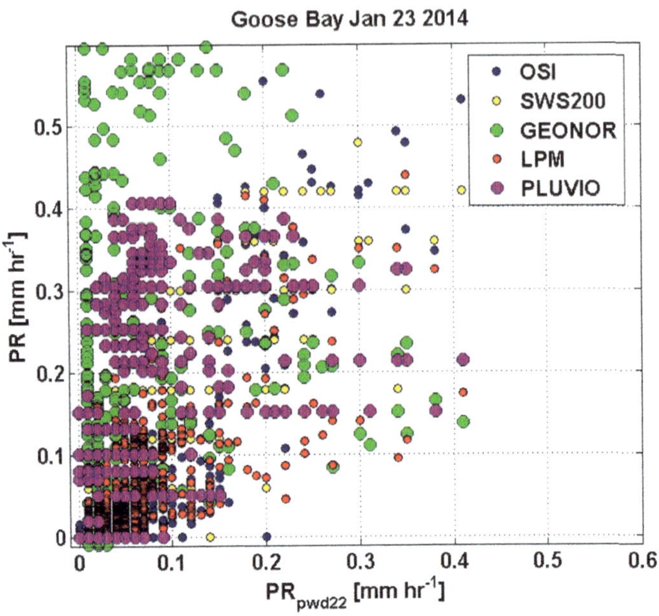

Figure 4 The scatter plot of precipitation rate (PR) obtained from four sensors (Table 1) against PR obtained from PWD22 during a light snow event occurred at T = −18°C over Goose Bay on January 23, 2014.

PR, if Geonor is excluded, was about 0.6 mm/h. The scattering of PR values indicate that these sensors were highly sensitive to varying degree of shape, size, and number concentration of ice crystals. Increasing LSN PR values result in large PR scattering for all sensors. Both LPM and OSI PR against PWD PR show less scattering compared to other PR measurements. The Pluvio-based PR values were in better agreement with optical sensor PR values compared to the Geonor unshielded PR measurements.

The LPM sensor described above is also used for obtaining snow particle spectral characteristics (Fig. 5). The LPM can accurately measure sizes greater than 500 μm. In this figure, the y-axis represents particle fall velocity (V_f), the x-axis represents particle diameter (L), and the color bar indicates the number of ice crystals (Fig. 5A). This plot shows that at 12:30 UTC, the maximum PR was 0.45 mm/h (Fig. 5B) representing light snow conditions (<0.5 mm/h). It is seen from Fig. 5A that most of the particles during light snow events have size ranges less than 600 μm. Under windy conditions, these small particles can move with wind speed, therefore V_f can be more than a typical ice crystal fall speed occurring in the calm weather conditions.

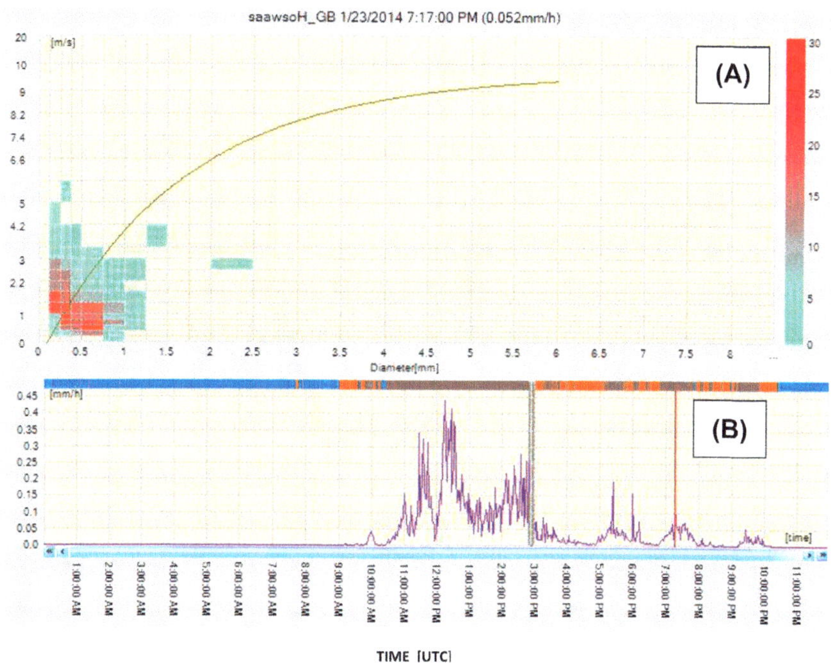

Figure 5 The laser precipitation monitor (LPM) measured particle spectra (A) during a light snow event on January 23, 2014 occurred over Goose Bay, NFL. y axis is terminal velocity, x axis for the diameter, *color bar* for the particle counts, and *solid line* for the theoretical droplet fall velocities. Precipitation rate time series is shown on (B). After 16:00 UTC (coordinated universal time) (4 p.m.), LPM reacted for blowing snow particles.

The most important impact of light snow, including ice fog deposition, on weather is due to its reduced visibility through increasing extinction of radiation. In order to show this effect, time series of visibility obtained by five sensors given in Table 1 are plotted in Fig. 6. Both FD12p visibility and PWD visibility were consistent with each other. The OSI visibility is much noisier compared to other sensors. Its measurements changed abruptly sometimes (e.g., at 13:00 UTC). The PWD, Sentry, and SWS visibility were usually consistent with each other but Sentry visibility had relatively lower values than the others instruments.

The visibility scatter plot suggests that uncertainty in visibility from any sensor, relative to FD12p visibility, can be as high as 2 km for visibility <5 km (Fig. 7). This uncertainty can be more than 5 km when FD12p visibility is larger than 10 km. The lowest uncertainty is obtained with the Sentry sensor and the highest uncertainty results from the use of the OSI sensor. The red lines on this plot show confident intervals for 10% and

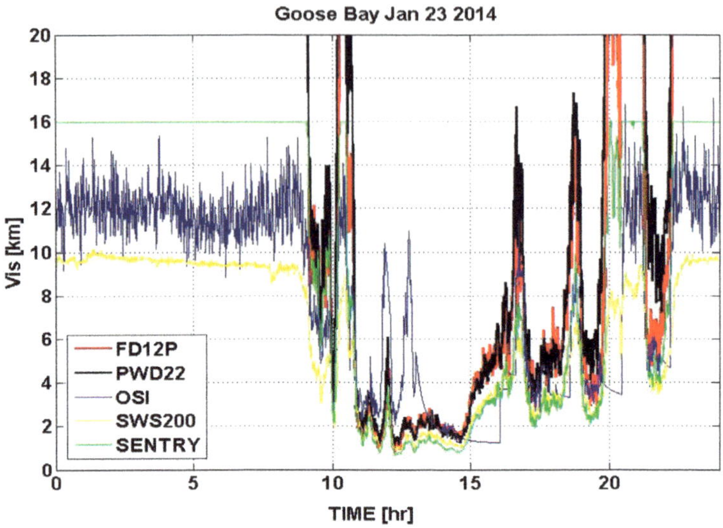

Figure 6 The visibility time series in UTC (coordinated universal time) obtained from various sensors shown in Table 1 for January 23, 2014 light snow case (10:00–16:00 UTC) occurred at about −18°C. After 16:00 UTC, blowing snow effects on sensors are seen.

90% percentiles around the mean values (large red circles) when visibility from Sentry is plotted against FD12p visibility. The FD12p visibility is usually 25% larger than Sentry visibility. Large scattering on this plot likely indicates the optical sensitivity of the sensors to ice crystals and algorithms used for visibility calculations.

Measurements of particles from both the LPM and GCIP instruments can be integrated into a single size spectrum (from 10 μm up to 1 cm size range). This spectrum can be used in validation of numerical weather prediction (NWP)-based PR and PA simulations, and remote sensing retrievals (not presented here). The GCIP and LPM sensors have a better sensitivity for size distributions for sizes lower and higher than about 500 μm, respectively. This can allow researchers to calculate deterministic parameters for microphysical parameterizations that include visibility, N_i, and Ice Water Content (IWC). These results can also be used for postprocessing of model-based simulations (Gultepe et al., 2015; Milbrandt & Yau, 2005; Morrison, Curry, & Khvorostyanov, 2005). Fig. 8 shows some examples of LSN particles obtained using the GCIP instrument that were mostly single particles e.g., plates, dendrites, and columns. Some large particles from the GCIP instrument were observed, but they were not common during

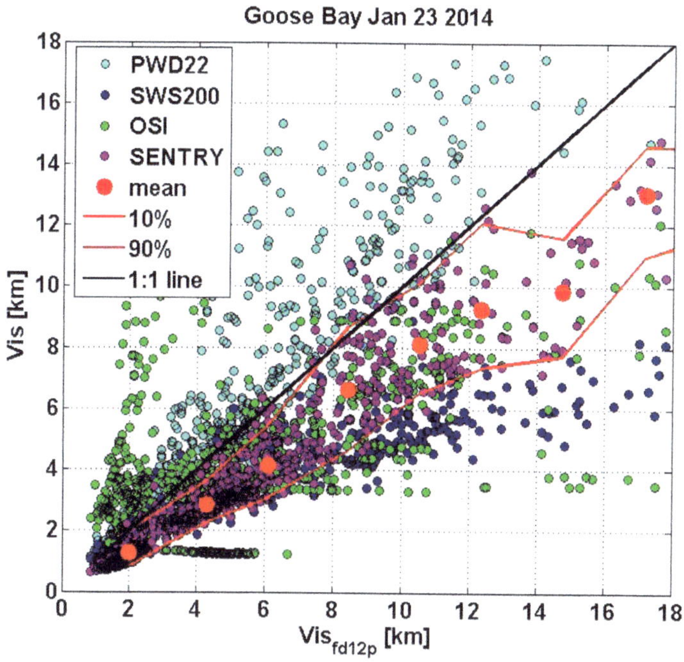

Figure 7 The visibility scatter plot of the measurements obtained from four sensors versus FD12p for January 23, 2014 case. The *solid black line* is for 1:1 line. The *solid red lines* (10% and 90% confident intervals) are for Sentry visibility versus FD12p visibility obtained over 2 km visibility intervals over x axis. The *filled red circles* are for binned averaged visibility values.

LSN conditions. Ice crystal sizes shown on Fig. 8 were usually less than 800 μm. The purpose of this figure is to emphasize importance of ice crystal shape and density [indicated with colors of yellow (25% of shadowing), red (50%), and black (75%)] that can be used for estimation of LSN PR from the GCIP as shown in Fig. 9.

The GCIP, LPM PR, Pluvio, and PWD measurements are shown in Fig. 9. The solid green colored line is for 1 min averages of GCIP PR that is based on 1 Hz observations (black dots). The PR from GCIP is obtained as

$$\mathrm{PR_{GCIP}(mm/h)} = \mathrm{A_c} \sum_{7.5\ \mu\mathrm{m}}^{930\ \mu\mathrm{m}} \mathrm{V_i(D)}\rho_s\mathrm{(D)N_i(D)V_f(D)}\,, \tag{3}$$

where $\mathrm{V_i(D)}$ ($\mathrm{cm^{-3}}$) is the snow crystal volume for particle diameter (D). $\mathrm{PR_{GCIP}}$, using empirical relationships between mass and size, is obtained

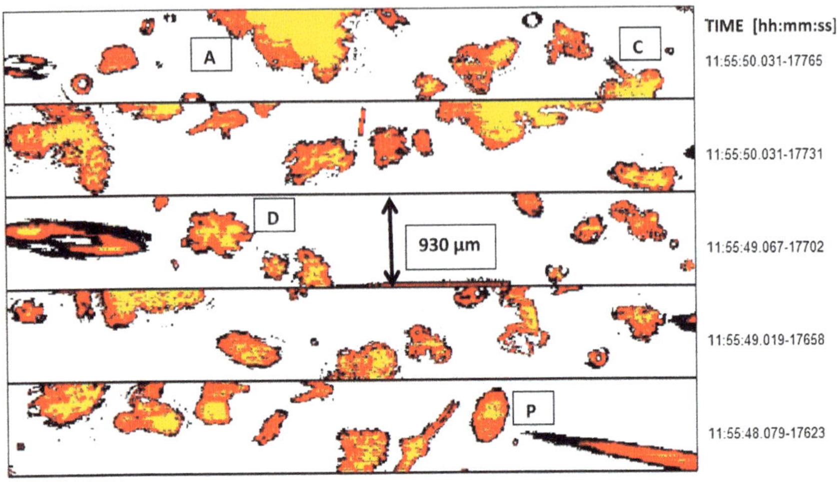

Figure 8 Some examples of light snow (LSN) particles (e.g., ice crystals) obtained using ground cloud imaging probe instrument on January 23, 2014 that were mostly single particles e.g., plates (P), dendrites (D), and columns (C), and some aggregates(A). Maximum ice crystal size for this sensor is less than 930 μm. LSN crystals have various shapes and densities [indicated with colors of yellow (25% of shadowing), red (50%), and black (75%)]. Time scale on the right side of the box is given by hh:mm:ss and total time required to fill the buffer with ice crystals in microseconds.

from the known particle spectra with bins of $\Delta D = 15$ μm. In Eq. (3), ice crystal mass, $m(D) = V_i(D)\rho_s(D)$, is the function of particle shape. Overall, results showing the location of the maximum and minimum PR values are comparable to these of LPM but large differences exist in the absolute values. The GCIP maximum values are usually less than LPM maximum values. This is likely due to upper threshold (930 μm) of the GCIP compared to that of LPM (e.g., 1 cm). Some sections of GCIP PR measurements show PR < 0.1 mm/h (e.g., between 05:00 and 06:00 UTC) during which both LPM and PWD PR were absent. This result is explained by the sensitivity of GCIP probe for small single ice crystals and undercatch issues related to optical probes. The Pluvio-based PR was usually significantly larger (30%) compared to PRs from other sensors.

4.2 May 3, 2014 Light Snow Event

This section is organized similar to the January 23 case. On this day, LSN PR and PA were usually less than 1 mm/h and 1.5 mm, respectively, and T was about 0°C. Wind speed was usually less than about 4 m/s during the snow

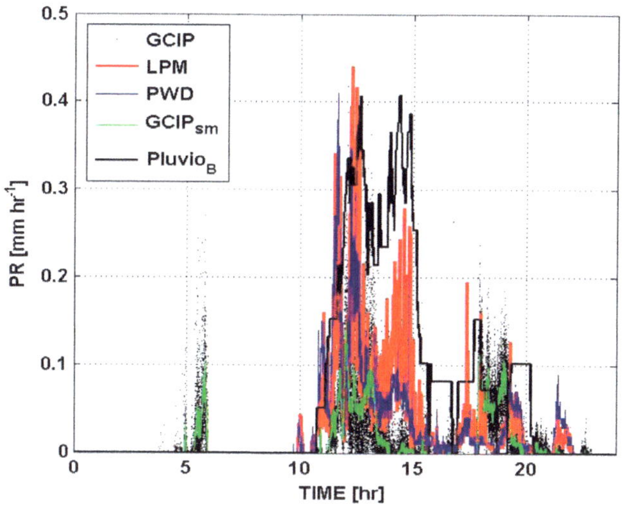

Figure 9 Time series in UTC (coordinated universal time) of ground cloud imaging probe (GCIP), laser precipitation monitor (LPM), Pluvio, and present weather sensor (PWD) based light snow precipitation rate (PR) on January 23, 2014, Goose Bay. The *black dots* are for 1 Hz PR obtained from GCIP. The *green solid line* is for 60 s averages of GCIP PR to match with LPM and PWD based PR scales. The Pluvio-based PR is obtained using 60 min running averages.

event, indicating likely negligible blowing snow conditions when snow crystals are wet. This case was significantly different from the January 23 case which had much higher T conditions. However, both events had relatively similar wind conditions.

4.2.1 Profiling Microwave Radiometer Observations

Fig. 10 shows PMWR time-height temperature (box A) and relative humidity with respect to ice (box B) cross-sections to 1.2 km height, and liquid water content (box C). Also shown are IWV and ILW retrievals (box D), and Tamb and Tir (box E).

Lower diurnal variability as large as 8°C is seen in the temperature cross-section. The peak value occurs near 20:00 UTC (17:00 local time). Inversions dominate the RH_i cross-section, peaking near 600 m height from 07:00 to 13:00 UTC. These are intermittent at other times. Liquid water content near 600 m height peaks near 6 g/m^3 at 08:00 UTC, and again at slightly higher density near 18:00 UTC. Clear air is evident from 00:00 to 02:00 UTC, with moderate clearing near 06:00, 14:00 and 23:00 UTC.

IWV gradually increases from 0.6 to 1.4 g/m^3 from 00:00 to 18:00 UTC and then decreases to 0.9 g/m^3 by the end of the day. ILW shows a

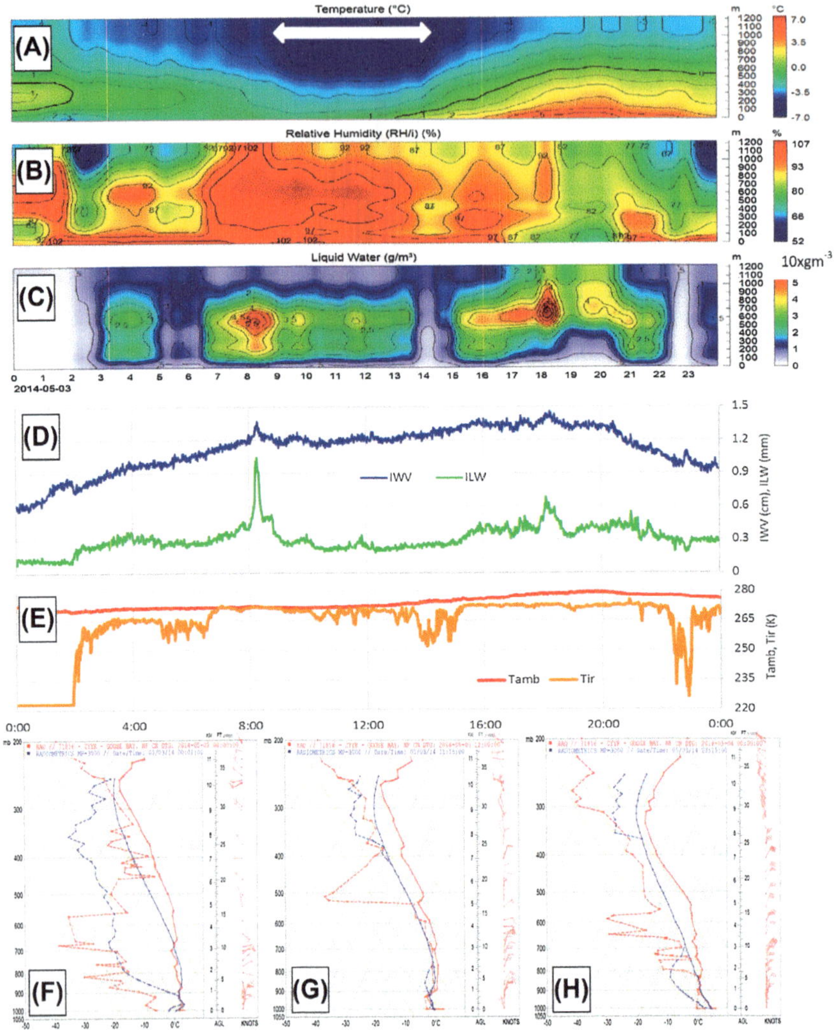

Figure 10 Radiometrics profiling microwave radiometer (PMWR) temperature (A), relative humidity over ice (B), and liquid water content (×10) profile retrievals to 1.2 km height (C), integrated water vapor (IWV) and integrated liquid water (ILW) retrievals (D), surface temperature (Tamb) and cloud base infrared temperature (Tir) (E), and Goose Bay radiosondes and PMWR profiles at 00Z, 12Z, 24Z are shown in (F–H), respectively, for May 3, 2014.

minimum ~0.15 mm from 00:00 to 02:00 UTC, corresponding to a flat, minimum T_{ir} value of 223K (no cloud). This indicates a 0.15 mm ILW bias probably resulting from calibration accuracy degradation and NN training with distant radiosondes (>1000 km). T_{amb} and T_{ir} in box E show LSN (T_{amb} ~ T_{ir}) or low cloud after 02:00 UTC for most of the day.

Radiosonde and PMWR T soundings up to 1 km height and Td soundings up to 3 km height, representing at 00:00, 12:00, 24:00 UTC (boxes F, G, and H, respectively). These profiles show good agreement within the cloudy layers, in particular when considering the highly variable radiosonde Td measurements seen at 00:00 and 24:00 UTC. Again, upper level PMWR retrievals show systematic bias consistent with NN's trained with distant radiosondes and possible calibration accuracy degradation. In this case, LWC values were relatively larger than those of January 23, 2014 case. However, LWC values decreased significantly during the LSN event.

4.2.2 In Situ Observations

Time series of PR from various sensors are shown in Fig. 11A. Sensors showed significantly different responses for LSN PR. The worst response to light snow conditions was from both the Geonor and FD12p instruments (box A). FD12p values were occasionally higher than those of others. This may be due to condensation/deicing issues. The Geonor observations resulted in overestimation of PR which was likely due to wind effects on transducers. Its measurements were occurring randomly even in the absence of precipitation. The Pluvio measurements, as a reference platform, were consistent with most of the optical sensors. The FD12p PR was relatively larger compared to PR from the other optical sensor. LPM measurements were usually consistent with PR obtained from PWD, OSI, and SWS. However, both OSI and SWS occasionally measured larger PR values. T was about −2°C during the event but reached 7°C during the end of the period shown (box B). Uh was up to 4 m/s during event (box C). The unexpected spikes in the sensors after snow event, especially for Geonor, were most likely a result of blowing snow and gust conditions.

A scatter plot on PR for this case is shown in Fig. 12. This figure suggests that the maximum PR was about 1 mm/h. The scattering of PR values is consistent with the January 23 case. It seems that LPM PR is better related to PWD PR compared to other measurements. Both the OSI and Geonor PR were much larger than the PR values of other sensors. The Pluvio PR was also found to be comparable with PWD PR.

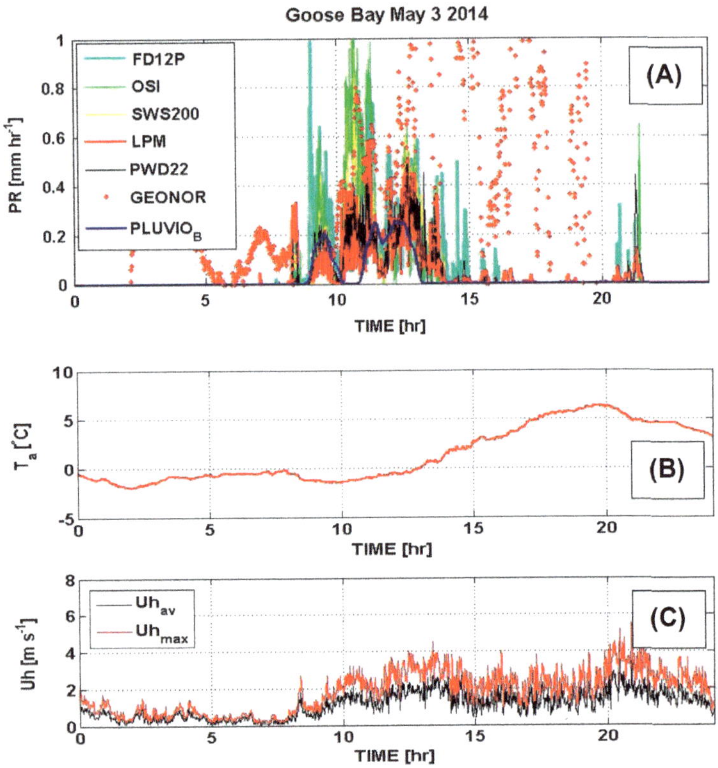

Figure 11 The comparison of light snow precipitation rate from various instruments (see legend, also in Table 1) at T = −2°C occurred over Goose Bay on May 3, 2014 are shown in box (A). T and Uh observations from WXT sensor over 1-min intervals (averaged and maximum) are shown in box (B and C), respectively.

The snow particle spectral characteristics and PR obtained from the LPM measurements are shown in Fig. 13. This plot shows that at 11:15 UTC, the maximum PR was ~0.5 mm/h (Fig. 13B). Snow crystals during the event had size ranges usually less than 1000 μm with V_f < 1.2 m/s. Therefore, influence of blowing snow on PR was likely negligible. These results are consistent with nonblowing snow conditions when Uh < 7 m/s (Trouvilliez et al., 2015).

The time series of visibility, similar to Fig. 6, are shown in Fig. 14. All visibility values were reasonably similar, but the variability was large and reached up to 2 km when visibility = 2 km. This was similar to January 23 case. The Sentry visibility was usually small compared to other measured visibility values. However, the PWD visibility was usually larger than visibility measured from the other sensors.

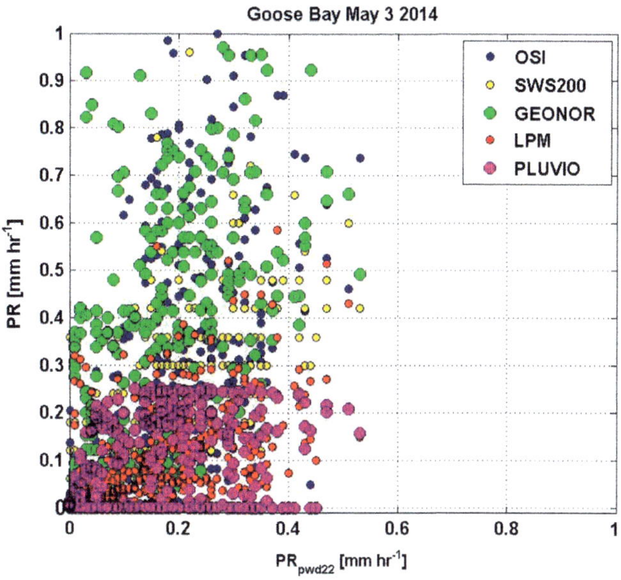

Figure 12 The scatter plot of precipitation rate (PR) obtained from four sensors (Table 1) against PR obtained from present weather sensor 22 (PWD22) during a light snow event on May 3, 2014 at T = −2°C occurred over Goose Bay on May 3, 2014.

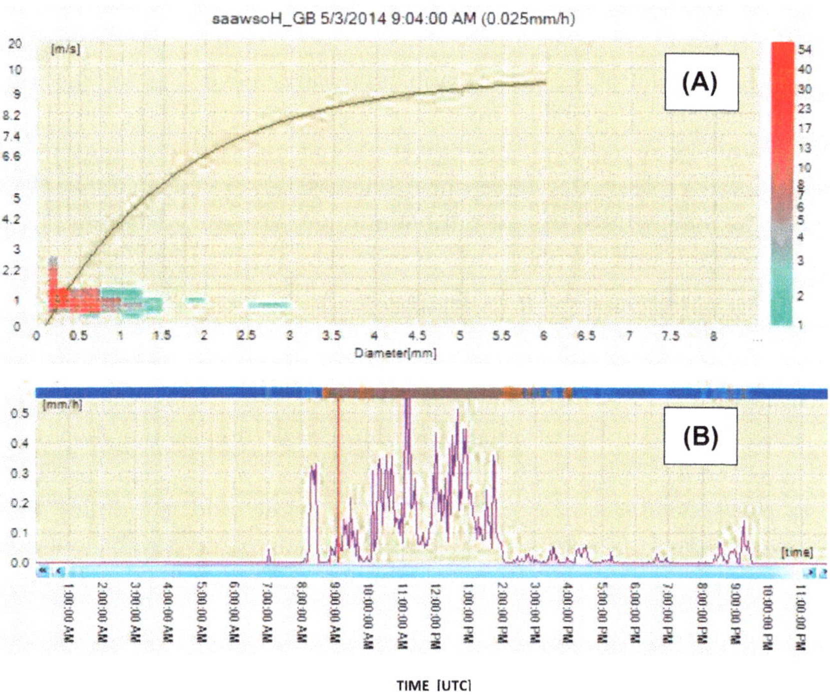

Figure 13 The laser precipitation monitor measured particle spectra (A) during a light snow event on May 3, 2014 occurred over Goose Bay. y axis is the terminal velocity, x axis for the diameter, *color bar* for the particle counts, and *solid line* for the theoretical droplet fall velocities. Precipitation rate time series in UTC (coordinated universal time) is shown on x axis (B).

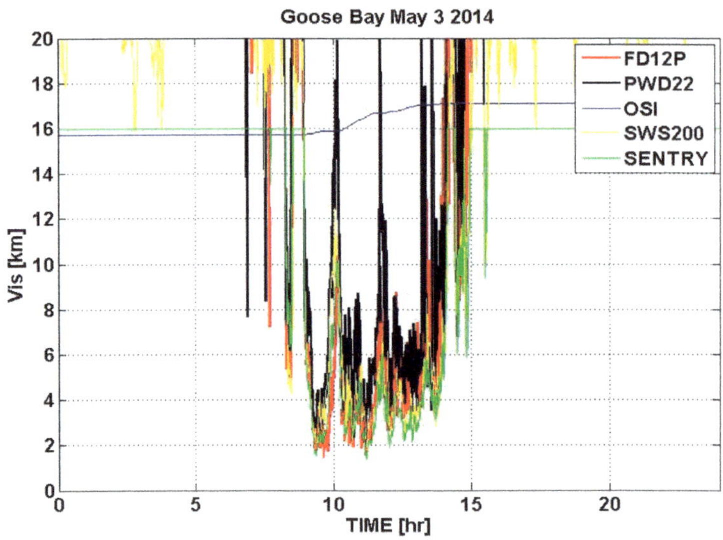

Figure 14 The visibility time series in coordinated universal time obtained from five sensors shown in Table 1 for May 3, 2014 light snow case occurred at about −2°C over Goose Bay on May 3, 2014.

The scatter plot of the various visibility measurements (Fig. 15) suggests that uncertainty in visibility from any sensor compared to a reference platform can be up to 100%. Icing or wet LSN accumulated on some sensors likely played a role on this discrepancy. The OSI sensor was usually iced up and its measurements were not reliable during wet snow conditions. This was a common issue for the entire project. The lowest values of visibility are usually obtained with the Sentry sensor, while the highest values were obtained with the PWD and SWS sensors. The red lines on this plot show confidence intervals. The red filled circles are for bin-averaged visibility values.

Examples of LSN particles and freezing drizzle droplets are shown in Figs. 16 and 17, respectively. Snow particles were mostly single particles (e.g., plates, dendrites, and columns, Fig. 16) while moderate riming was observed. Some large particles (>900 μm) were also observed. The freezing drizzle occurred at about 18:45 UTC (Fig. 17) during which surface T was less than 0°C (icing occurred on sensors) and droplet sizes were generally less than 500 μm at this time. This was one of the reasons for the discrepancy in visibility among the sensors when freezing precipitation occurred. The T at 18:00 UTC was about 5°C when freezing drizzle occurred.

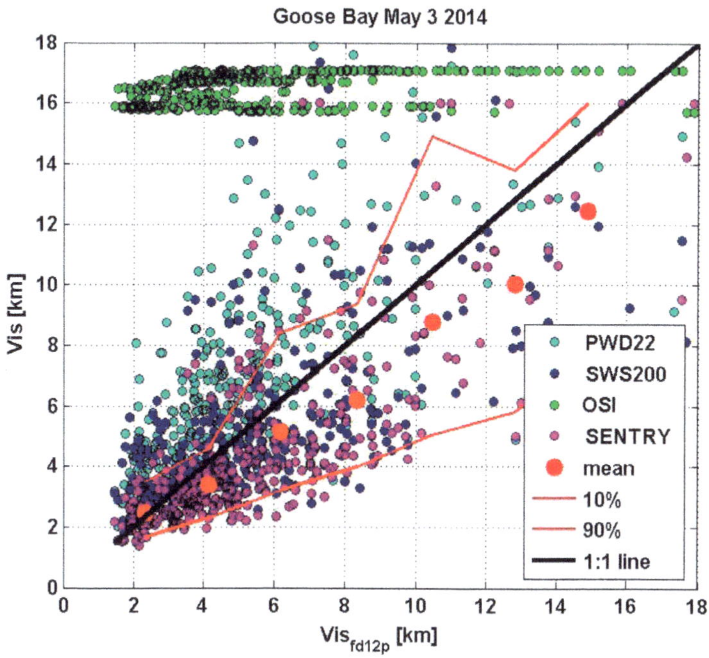

Figure 15 The visibility scatter plot of the measurements obtained from four sensors versus FD12p for the May 3, 2014 case. The *solid black line* is for 1:1 line. The *solid red lines* (10% and 90% confident intervals) are for Sentry visibility versus FD12p visibility obtained over 2 km visibility intervals over x axis. The *filled red circles* are for binned averaged visibility values.

Time series of GCIP, LPR, PWD, and Pluvio PR are shown in Fig. 17. The Geonor observations are not shown here because of their relatively unexpected large values likely because of wind conditions. The large PR differences here in GCIP and others optical sensors were likely due to particle mass-size relationships used. The Pluvio PR was usually found to be comparable with LPM PR.

4.3 Variability Issues

The variability of light snow precipitation measurements can be large compared to heavy snow. In order to quantify this uncertainty, both LSN events are separately analyzed for all precipitation conditions assuming that the PWD sensor indicates occurrence of precipitation for PR $>$ 0.01 mm/h. The Pluvio measurements are used in the analysis as reference platform. Table 2 (3) and Table 4 (5) show PR (visibility) mean,

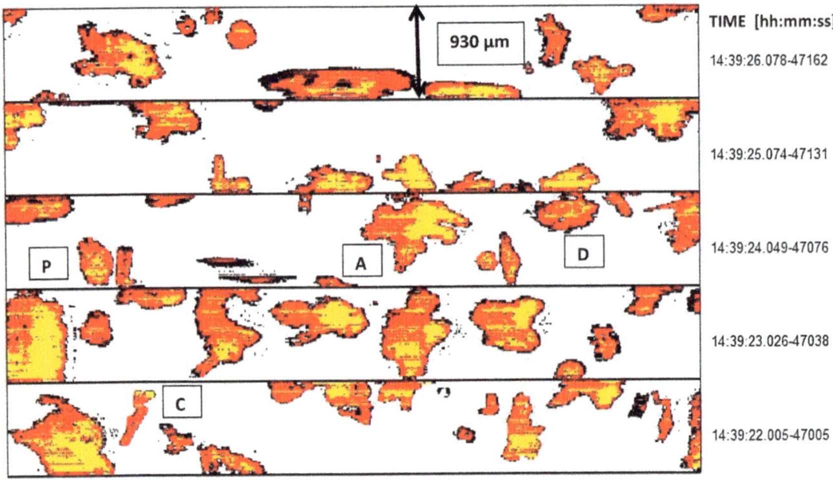

Figure 16 Some examples of light snow (LSN) particles (e.g., ice crystals) obtained using the ground cloud imaging probe instrument on May 3, 2014 that were mostly single particles e.g., plates (P), dendrites (D), and columns (C), and some aggregates(A). Maximum ice crystal size for this sensor is less than 930 μm. LSN crystals have various shapes and densities [indicated with colors of yellow (25% of shadowing), red (50%), and black (75%)]. Time scale on the right side of the box is given by hh:mm:ss and total time required to fill the buffer with ice crystals in microseconds.

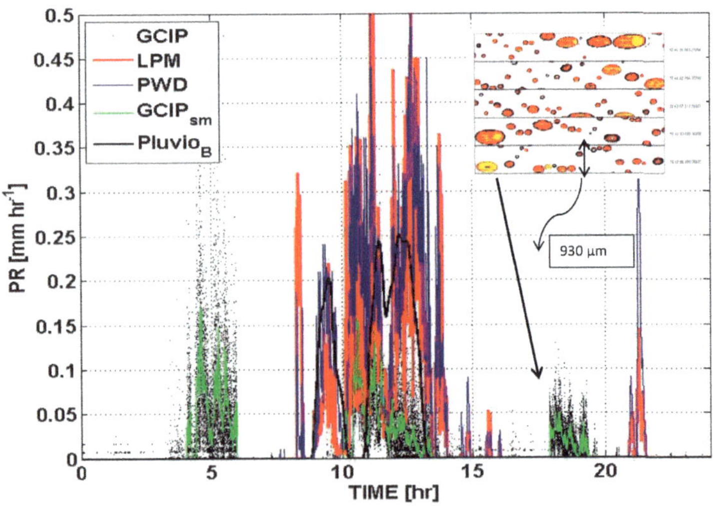

Figure 17 Time series in UTC (coordinated universal time) of ground cloud imaging probe (GCIP), laser precipitation monitor (LPM), Pluvio, and present weather sensor (PWD) based light snow precipitation rate (PR) on May 3, 2014, Goose Bay, Newfoundland. The *black dots* are for 1 Hz PR obtained from GCIP. The *green solid line* is for 60 s averages of GCIP PR to match with LPM and PWD based PR scales. The Pluvio-based PR is obtained using 60 min running averages. Freezing drizzle droplets occurred at 18:30 UTC is seen in the inset box. Note that comparisons did not include segments with droplets.

Table 2 Shows mean precipitation rate (PR), sd, and variability (sd/mean) between PWD22 PR (PR_{pwd}) and other sensor visibility for January 23, 2014

Instrument	Mean PR (mm/h)	Sd PR (mm/h)	Variability	ΔPR (mm/h)
PWD22	0.07	0.07	1.00	Reference
Geonor	0.89	0.88	0.98	+0.82
LPM	0.08	0.08	1.00	+0.01
OSI	0.06	0.11	1.83	−0.01
SWS	0.10	0.10	1.00	+0.03
GCIP	0.06	0.05	0.95	−0.01
Pluvio	0.17	0.13	0.77	+0.10

It is assumed that $PR_{pwd} > 0.01$ mm/h. Because of its sensitivity to LSN, PWD is chosen as a reference platform.

Table 3 Shows mean visibility (Vis), sd, and variability (sd/mean) between PWD22 visibility and other sensor visibility for January 23, 2014

Instrument	Mean Vis (km)	Sd Vis (km)	Variability	ΔVis (km)
PWD22	4.42	4.01	0.90	Reference
FD12p	3.45	2.36	0.68	−0.97
SWS	2.43	1.37	0.56	−1.99
Sentry	2.32	1.72	0.74	−2.10
OSI	3.77	2.09	0.55	−0.65

It is assumed that $PR_{pwd} > 0.01$ mm/h.

Table 4 Shows mean precipitation rate (PR), sd, and variability (sd/mean) between PWD22 PR (PR_{pwd}) and other sensor visibility for May 3, 2014

Instrument	Mean PR (mm/h)	Sd PR (mm/h)	Variability	ΔPR (mm/h)
PWD22	0.17	0.11	0.65	Reference
Geonor	0.35	0.77	2.20	+0.28
LPM	0.12	0.11	0.92	−0.05
OSI	0.24	0.27	1.13	+0.07
SWS	0.19	0.17	0.90	+0.02
GCIP	0.11	0.09	0.88	−0.06
Pluvio	0.10	0.09	0.99	−0.07

It is assumed that $PR_{pwd} > 0.01$ mm/h. Because of its sensitivity to LSN, PWD is chosen as a reference platform.

standard deviation (sd), variability (sd/mean), and absolute difference between Pluvio and others sensors for January 23 and May 3, 2014, respectively.

Variability in PR time series for the January 23 case was up to 100% (Table 2). visibility variability was also close to 100%. Less variability in

Table 5 Shows mean visibility, sd, and variability (sd/mean) between PWD22 visibility and other sensor visibility for May 3, 2014

Instrument	Mean Vis (km)	Sd Vis (km)	Variability	ΔVis (km)
PWD22	8.77	5.63	0.64	Reference
FD12p	9.85	13.64	1.39	+1.08
SWS	6.40	5.02	0.78	−2.37
Sentry	5.33	4.35	0.82	−3.44
OSI	16.54	0.5	Iced up	+7.77

It is assumed that $PR_{pwd} > 0.01$ mm/h.

visibility is seen in the SWS and OSI measurements (Table 3). Both Tables 4 and 5 for the May 3 case also suggested similar variability results for PR and visibility. Icing on the May 3 case caused severe problems in the OSI measurements compared to other sensors despite that they have all heating units. Therefore, OSI should be used cautiously under icing conditions.

Comparisons of PR derived from the GCIP and LPM versus PWD (Fig. 9 for January 23 and Fig. 17 for May 3) suggest that the averaged ΔPR is usually less than 0.06 mm/h (up to 50%, defined as ΔPR/mean) (Table 2). Geonor ΔPR values were much larger than other PR measurements. The Pluvio PR values were relatively larger than those of optical sensors for January 23 case but they were smaller for May 3 case. GCIP measurements were usually comparable to PR from other sensors. During the project, observations indicated that GCIP responded very well for light snow and ice crystals, unless there were very gusty wind conditions. Because of the intake of air at 25 m/s into the sampling area, GCIP was not affected by the large range of wind speeds. Note that differences between PRs from optical probes can be related to sensor algorithm and calibration, and lower threshold values used for detecting precipitation.

The visibility measurements from the present weather sensors (Figs 6 and 14) are shown in Tables 3 and 5 for January 23 and May 3, respectively. The PWD visibility is used as reference because of its response to LSN conditions. For the January 23 case, visibility differences between PWD and others sensors are estimated to be about at least 50% but this value decreased significantly for May 3 case, excluding for OSI visibility (Table 5). These suggest that optical sensors for visibility measurements need to be improved and calibrated for cold weather conditions. Note that OSI was performing worst among all sensors on May 3 because icing usually formed over the sensor lenses (see green filled circles upper end of Fig. 15).

Table 6 Precipitation amounts (PAs) are shown for various sensors for each light snow case

Cases	January 23, 2014	May 3, 2014
Sensor	PA(ΔP) mm	PA(ΔP) mm
Pluvio (Reference)	1.50(0.00)	0.55(0.00)
FD12p	1.35(−0.15)	1.35(+0.80)
OSI	0.35(−1.11)	1.25(+0.70)
SWS	0.65(−0.85)	1.15(+0.60)
LPM	0.55(−0.95)	0.65(+0.10)
PWD	0.50(−1.00)	0.85(+0.30)
GCIP	1.00(−0.50)	0.85(+0.30)
Geonor	4.00(+2.50)	4.00(+3.45)

ΔP is difference between sensor PA and PA from Pluvio at the end of each precipitation event. To check sensor sensitivity, Pluvio was used as reference platform here. Note that difference among all sensors was less than 1 mm of accumulated snow amount, suggesting large variability and uncertainty.

Based on PAs from all optical and weighing sensors, Table 6 suggests that PA uncertainty can be more than 50%. The PA from Pluvio as a reference platform was usually larger than these of optical sensors for January 23 case, and lower than these of May 3 LSN case. Geonor based PA values using two or three vibrating wires were usually either overestimating or underestimating PA compared to other sensors. Optical probes with lower sensitivity for PR measurements by detecting small particles were usually consistent each other (Table 6). Overall, variability in PA and visibility from various sensors was up to 50−100% for LSN conditions.

5. DISCUSSIONS

Discussions are divided into the five subsections. In these subsections, issues related to LSN measurements, instruments, forecasting/nowcasting and aviation, and climate change assessment and averaging scales are discussed.

5.1 Snow Measurements

Snow precipitation can be measured by in situ sensors including weighing gauges, present weather observing systems (PWOS), optical spectra precipitation probes (OSPPs), and by remote sensing methods such as satellite and radar based retrievals, indirectly. Details on snow measurements and its relation to the energy concepts can be found in Rudolf and Rubel (2005a, 2005b). Traditional in situ snow measurements on the ground can be made by measuring snow depth and weight, and then estimating snow water equivalent (SWE). Water equivalent rate of snow precipitation can be obtained using weighing gauges if the amount of snow falls into their measuring volume pass a certain threshold which is usually in the range of >0.1–0.5 mm/h (Rasmussen et al., 2012). The sensitivity of the PWOS for snow measurement is <0.01 mm/h (Table 1). In situ measurements of snow precipitation using weighing gauges cannot be accurate until wind corrections are performed and that is based on DFIR (Double Fenced Intercomparison Reference, Goodison and Yang, 1995) and/or bush gauge snow observations (Yang, 2014). However, these corrections cannot work for LSN conditions because of LSN particle density, shape, and crystal smaller sizes.

The LSN observations made from satellite and radar-based techniques may also have large uncertainties because of the nature of their retrieval methods (Tian, Peters-Lidard, & Eylander, 2010) and higher thresholds (0.5–1 mm/h) for detecting accurate signal for snow precipitation. Clearly, the snow precipitation amount and rates obtained from remote sensing techniques are based on assumptions such as particle spectra that likely vary with meteorological conditions. Therefore, the assumption regarding fixed particle spectra for satellite retrievals needs to be improved for cloud and snow studies.

Complex shields designed for snow weighing instruments (Smith, 2009) are not always effective because winds often have a vertical component and can be very chaotic in frontal systems (Gultepe et al., 2014b, 2014a). Therefore, wind corrections applied for snow measurements cannot work for all snow types and locations in the same way. The remote sensing techniques for snow retrievals using satellite observations can provide a better areal coverage compared to in situ sensors (Huffman, Adler, Rudolf, Schneider, & Keehn, 1995; Rudolf, Hauschild, Rüth, & Schneider, 1994, 1996). Unfortunately, these techniques cannot work for LSN accurately because of low PR for LSN (between 0.01 and 0.5 mm/h). Rudolf and Rubel (2005a, 2005b) suggested that 0.2 mm/d (5 W/m^2) is a measure of the accumulated uncertainties in the independently estimated parts of the global annual water budget. This is two times more than the precipitation trend calculated for last 100 years. Certainly, averaging scales are important for PR calculations but the uncertainty of LSN measurements complicates the estimates. If LSN occurs at 0.5 mm/day, its occurrence then becomes a significant fraction of the total precipitation for the regions with PR less than 50 mm/year.

Rudolf and Rubel (2005a, 2005b) stated that annual PR over land is 2.09 mm/d. The daily missing LSN PR estimated at about 0.2–1 mm/d in the present work can be very important for Arctic and continental cold climate budget studies. This amount can reduce the climate warming significantly. This is refereed here as *direct effect of light snow* (DE-LSN) on the radiation budget (up to -30 W/m^2). The LSN results in decreasing atmospheric vapor by cleaning of the atmosphere at higher levels and this leads to increasing IR cooling at the surface (up to 30 W/m^2). Here, this effect is called the *indirect effect of light snow* (IE-LSN) on the radiation budget. Both effects lead to cooling at the surface because of loss of energy to the space (Girard & Blanchet, 2001a, 2001b; Gultepe et al., 2009). These effects should not be mixed with snow-albedo feedbacks in the heat budget calculations.

5.2 In Situ Instrumental Issues

Instruments used for measuring LSN PR are not usually sensitive to low PR less than 0.5 mm/h (Gultepe et al., 2014a, 2014b; Rasmussen et al., 2012). In this category, weighing gauges (e.g., Geonor and Pluvio), and some OSPPs such as OTT and LPM disdrometers can be listed. These instruments can't measure snow precipitation rates accurately at very cold temperatures for LSN conditions. They are likely not calibrated and tested for. On the other hand, OPWSs sensitive to smaller particles (e.g., sizes less than 500 μm down to a few μm) can be used to obtain LSN PR. Clearly, OPWS and OSPPs are in need of further improvement for accurately measuring light snow in cold climates, especially when the particles are less than 500 μm. Zhang et al. (2015) stated that for $Uh > 3.5$ m/s, Geonor-based LSN measurements had very large errors with detecting LSN segments and its magnitude.

To show measurement issues related to LSN, Figs. 18 and 19 are provided for the time series of PA from various sensors and for Uh for January 23 and May 3, 2014, respectively. Geonor PA values are obtained using with two and three vibrating wires to test its sensitivity because it was providing very high readings. The Uh was about 2 ± 2 m/s for both days. Fig. 18A shows that using two wires, Geonor PA was much larger than others after about 2 h of snow start time at 10:00 UTC. When vibrating wire 2 is taken out, PA became more consistent with optical sensors after 14:00 UTC but the total PA (4 mm) was still much larger than the PA for all other sensors. Overall, notes taken during the data collection during project suggested that Geonor measurements were not reliable due to its response to the LSN

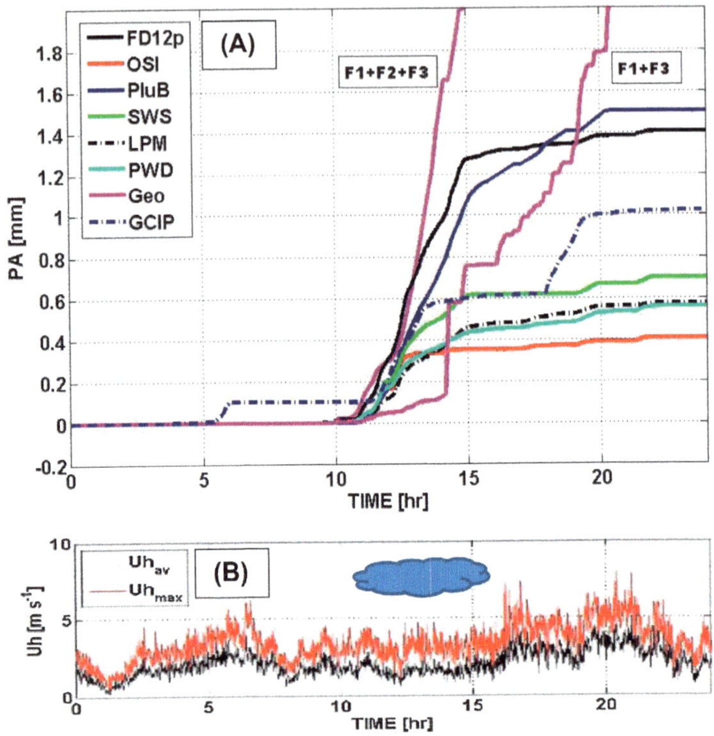

Figure 18 The accumulated precipitation amounts (PA) from eight sensors are shown in (A) and horizontal wind (Uh) from the WXT50 is shown in (B) for January 23, 2014. The F1, F2, and F3 indicate PA calculated using a specific vibrating wire frequency. Light snow segment is roughly indicated by a *cloud sketch*. These plots show that Geonor PA is very sensitive for response of vibrating wires to weight of accumulated snow and wind.

conditions (<0.5 mm/h). Pluvio PA as reference with single alter shield within a protected environment resulted in measuring comparable PA for January 23 but less for May 3. The GCIP PA was relatively larger as compared to other optical sensors in the later stages of the snow event. Fig. 18B shows that Uh was usually below 4 m/s, and its effect can be important for collection efficiency for weighing gauges. Pluvio PA was not affected significantly with this condition. Overall, LSN PA from various sensors ranged from 0.6 mm/h up to 1.6 mm/h.

Fig. 19A suggests that using three wires, Geonor PA was consistent with PAs from other sensors for the first 2 h after snow had started. Later it increased quickly to large values of PA up to 4 mm. PA estimated with two wires was found to be less than all others. The PA from Pluvio was

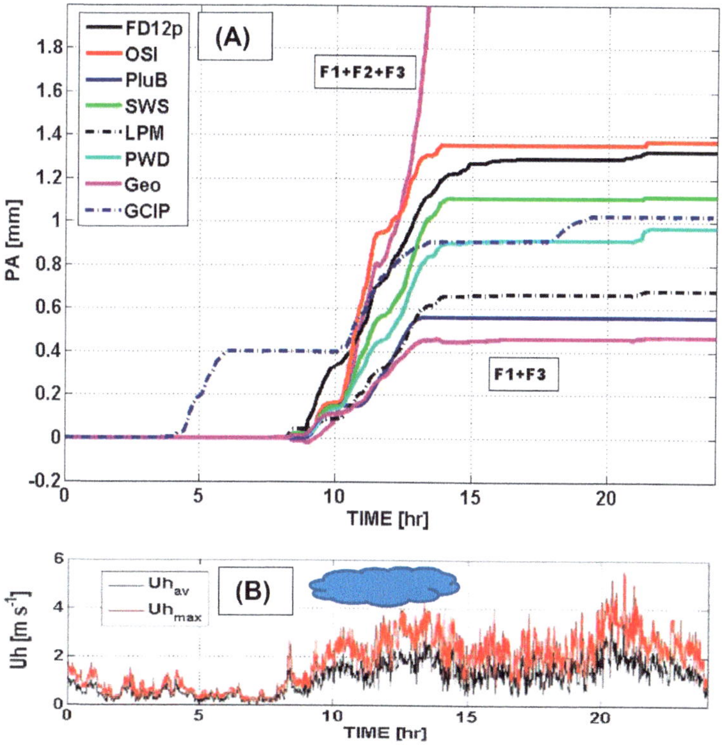

Figure 19 The accumulated precipitation amounts (PA) from eight sensors are shown in (A) and horizontal wind (Uh) from the WXT50 is shown in (B) for May 3, 2014. The F1, F2, and F3 indicate PA calculated using a specific vibrating wire frequency. Light snow segment is roughly indicated by a *cloud sketch*. These plots show that Geonor PA is very sensitive for response of vibrating wires to weight of accumulated snow and wind.

very close to it. The PA from LPM and GCIP usually agreed to each other. However, GCIP detected more snow segments for both cases. Overall, large variability was seen from optical sensors. The FD12p PA was usually larger than PA from other optical sensors. Note that 10% of SWE used in the PR algorithm determined by the manufacturers may not be valid for LSN calculations. During the LSN event for May 3, Uh was also similar to January 23 case with its magnitude less than about 4 m/s (Fig. 19B).

Overall, PA time series from the optical probes such as PWD and SWS are more consistent with each other compared to other sensors for both cases. Meanwhile, measurements from the weighing gauges are usually different than from those of PWD and SWS. Based on light wind conditions it is expected that weighing gauges and optical probes should have similar PA values. However, this is not the case. In the January 23 case, Pluvio measurements and those of optical probes are found to be comparable during early snow events but significant differences occurred at later times. In the May 3 case, PWD and SWS measurements were more consistent. However, weighing gauges were measuring either too low (likely due to wind induced errors) or too high PA values. For this case, OSI measurements were being affected by some icing. FD12p measured high PA values; however, PWD should measure PA better than FD12p because of new technology used in the PWD manufacturing. These results are consistent with Zhang et al. (2015) who stated that optical probes were more responsive for LSN conditions. Based on the above limited data set, use of optical sensors for LSN detection can only be used cautiously.

Based on the LSN cases studies presented here, weighing gauges may have significant issues for detection of LSN conditions that are related to the sensitivity of sensors and wind induced errors.

5.3 Snow Amount From Visibility Measurements

Snow amount and intensity can be obtained using visibility measurements, and it is commonly used for aviation applications. Visibility is usually obtained as a function of PR; visibility increases with decreasing PR. The visibility-PR relationship for LSN conditions defined here ($PR < 0.5$ mm/h) may not follow this inverse relationship. The United States National Weather Service (NWS) defines light snow when visibility >1 km. Rasmussen et al. (1999) used $PR < 1$ mm/h for LSN calculations. Under the LSN conditions, the NWS definition can be flawed because PR is not always indirectly related to visibility for $PR < 0.5$ mm/h. In fact there was a "*very light snow*" criteria in the past for the surface stations, but it was taken out to be reported. A direct relationship between light snow PR and visibility completely contradicts the concept used by aviation applications. Under the LSN conditions, both visibility and PR can be very small because of suspended ice crystals, resulting in a direct relationship (as seen in Fig. 20A, shown by a line with arrow).

The visibility-PR relationship representing the LSN conditions for SAAWSO ($Uh < 5$ m/s) and for heavy snow (HSN) conditions representing SAAWSO and FRAM projects (Fig. 20B) are given in Eqs. (4)–(6), respectively, as

$$\mathrm{Vis} = 0.9447\mathrm{PR}_{\mathrm{LSN}}^{-0.5136}, \mathrm{R} = -0.40, \tag{4}$$

$$\mathrm{Vis} = 0.7055\mathrm{PR}_{\mathrm{HSN}}^{-0.6957}, \mathrm{R} = -0.73 \tag{5}$$

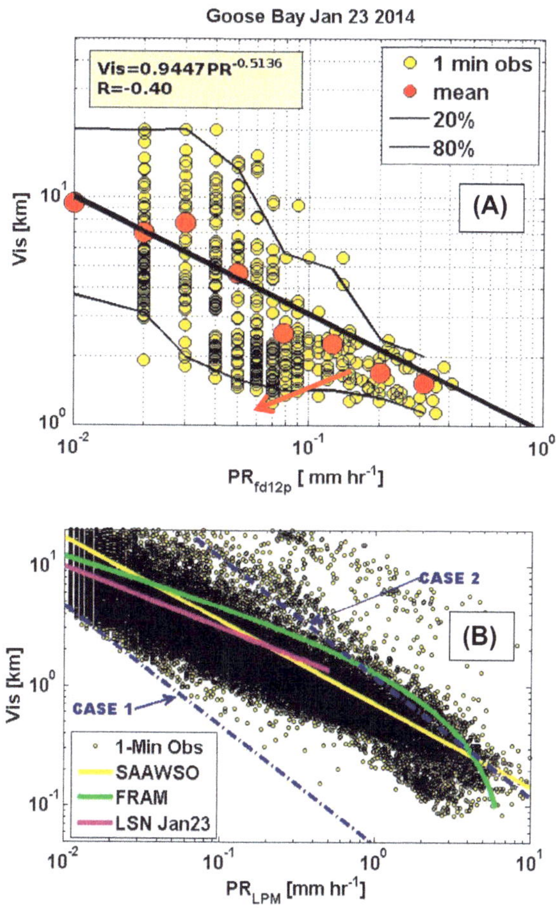

Figure 20 The 1-min observations of visibility versus precipitation rate (PR) obtained from present weather sensor (PWD) instrument during January 23, 2014 case (A). The *red solid line with arrow* indicates the region where visibility decreases with decreasing PR. PWD visibility versus laser precipitation monitor (LPM) PR for entire SAAWSO project is shown in (B). The *purple line* indicates the trend of mean values given in (A). The *yellow solid line* is a fit to measurements for entire satellite application for Arctic weather and SAR operations (SAAWSO) project data. For comparisons, visibility-PR relationship line obtained using FD12p sensor for entire FRAM project is shown with a *green line*. The *dashed blue lines* represent visibility and PR obtained from Eqs. (7)and(8), respectively. The Case 1 is for ice crystals with branches for low terminal velocities and the Case 2 is for small single solid ice crystals for higher terminal velocities.

$$\text{Vis} = 2.4319\text{PR}_{\text{HSN}}^{-0.3684} - 1.16. \ \text{R} = -0.65 \tag{6}$$

Comparisons suggest that for a given visibility (Vis) value, PR for FRAM is significantly larger than this of the SAAWSO project. This likely happens because of heavy and wet snow particles occurring at warmer and high moisture environments occurred during the FRAM project. For the LSN case, visibility-PR relationship (Eq. 4) with low correlation coefficient (R) suggests that visibility can be much smaller for a given PR value compared to intermediate snow conditions.

Assuming a gamma size distribution (Milbrandt & Yau, 2005; Rasmussen et al., 1999) visibility and PR equations can be obtained as

$$\text{Vis(cm)} = \text{A}\frac{\Gamma(m+1)(3.67+m)^2}{\Gamma(m+3)\text{D}_\text{o}^2\text{N}_{\text{ti}}}, \tag{7}$$

and

$$\text{PR(mm/h)} = \text{B}\rho_\text{i}\text{D}_\text{o}\text{V}_\text{t}/\text{Vis}, \tag{8}$$

respectively, where A and B are 2.49 and 4.68×10^4, N_{ti} the total ice crystal number concentration, m is the shape parameter, D_o median diameter, ρ_i ice crystal density, and V_t ice crystal terminal velocity. Using $N_{ti} = 0.001{:}0.01{:}10.0\ \text{cm}^{-3}$, $D_o = 1000\ \mu\text{m}$ (called as Case 1), with $\rho_i = 0.5\ \text{g/cm}^3$ and $V_t = 50$ cm/s for dry ice crystals with branches at cold temperatures. A visibility versus PR plot based on above equations is shown in Fig. 20B. Case 2 represents the same distribution with $D_o = 100\ \mu\text{m}$, $\rho_i = 0.1\ \text{g/cm}^3$, and $V_t = 100$ cm/s for dry solid ice crystals. The area between blue-dashed lines in Fig. 20B indicates that, during LSN conditions, small ice crystals with low densities can exist in large numbers. This condition may correlate directly with LSN PR. This suggests that visibility versus PR relationships should be evaluated cautiously for LSN conditions (Gultepe et al., 2015).

5.4 Statistics on Light Snow Measurements

Observations collected by the LPM sensor during the entire SAAWSO project are used to assess LSN impact on snow occurrence. Fig. 21 shows a probability density function (pdf) plot for PR over entire SAAWSO project that includes HSN conditions. The fit equation for the pdf of PR based on the Weibull distribution is obtained as:

$$\text{pdf} = 0.2\ \frac{\text{a}}{\text{b}}\left(\frac{\text{x}}{\text{a}}\right)^{\text{b}} e^{-\frac{\text{xb}}{\text{a}}} \tag{9}$$

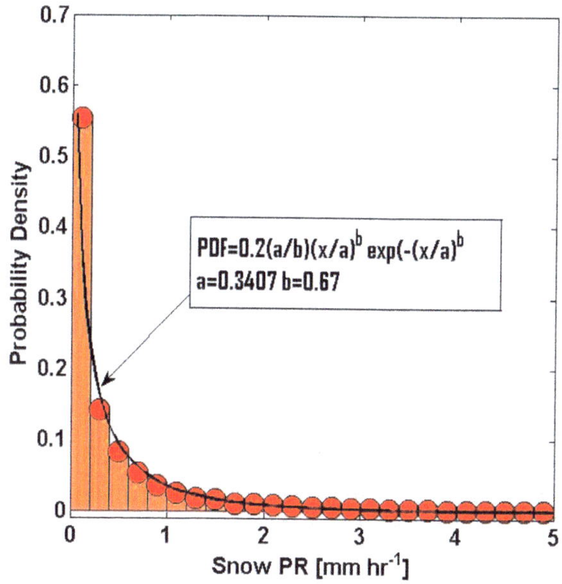

Figure 21 Shows histogram of precipitation rate (PR) for entire satellite application for Arctic weather and SAR operations project over Goose Bay from November 1, 2013 to May 1, 2014). The pdf of PR is obtained based on the Weibull distribution function given by Eq. (9) which is shown on the figure.

where x is the PR, $a = 0.3407$, $b = 0.67$, and 0.2 is the normalization factor for the fit. Fig. 21 suggests that PR (<0.5 (or 1.0 mm/h)) occurred at 75% (87%) of time, while the corresponding PA occurred at 11% (20%) of time. Based on observations, the LSN occurrence period was very large compared to heavy snow occurrence periods. Assuming that the Goose Bay area is receiving a total annual snow precipitation water equivalent amount of about 430 mm (http://www.currentresults.com/Weather/Canada/Newfoundland-Labrador/snowfall-annual-average.php), LSN PA/HSN PA is estimated to be 43 mm (11% of total snow precipitation occurrence). This ratio can be much larger over the cold and dry climates. For example, this ratio can be smaller over the marine and wet environments but it would be much larger over the Arctic continental regions. The LSN impact on climate change compared to HSN is likely much more important because of extinction of radiation over longer time periods. Note that the value of 43 mm is also almost equal to the entire Arctic continental tundra precipitation amount.

5.5 Missing Snow Effect in Climate Change Assessments

The LSN PA depends on how measurements are analyzed and averaged. The total amount of precipitation can change from a few 10s of mm per year to more than 1000 mm, depending on the location on the Earth. In the northern climates, the total snow precipitation amount is usually less than a few 100s of mm per year, and it is as low as 100 mm/year in the Arctic continental areas (Liston & Hiemstra, 2011). As shown in Ballinger et al. (2013), PA in Arctic tundra regions is almost 50% less than the one of other Arctic regions when the Köppen climate classification is used. The Köppen climate classification is based only on monthly averaged precipitations and temperatures, and it is provided with a code of 2- or 3-letters.

To assess the LSN impact on climate, PR should be related to its contribution to the energy budget by the process of sublimation. According to the energy conservation budget, a relationship between PR_{sn} (mm/day) and its equivalent energy amount Q_e (W/m^2) due to sublimation is found as

$$Q_e = K_c PR_{sn} L_{ice}, \tag{10}$$

where K_c is a conversion constant as 1/86,400 day, and L_{ice} (J/kg), is the latent heat of sublimation. When the surface is covered by snow, energy taken from surrounding air for evaporation of ice crystals (sublimation) is more than this required for surface covered by water (Barry, 1981). Using L_{ice} at 0°C as 2.83 MJ/kg and assuming PR = 1 mm/day occurring over the Arctic regions, Q_e becomes 32.8 W/m^2. This suggests that latent heat is released at cloud levels by condensation, and consumed by sublimation of snow crystals at the surface. Both effects modify the outgoing infrared radiative fluxes that results in net cooling at the surface. However, detailed analysis of LSN feedbacks using global climate models (GCMs) may be required to better assess its relationship on climate budget terms.

If 1 mm snow occurs daily with SWE amount as 30%; based on Eq. (10), this results in ~10 W/m^2 energy lost from the Earth surface to atmosphere. If we assume that LSN at about 1 mm/day occurs 20% of the time in Arctic regions over 6 month time period, this can result in up to a 2 W/m^2 uncertainty daily over a 6 month time period. This value is comparable to the direct and indirect aerosol effects given in the International Panel on Climate Change (IPCC) report (2007). Clearly, a daily LSN effect on local climate can be much larger than the seasonally averaged missing LSN effect. This missing LSN effect can further contribute to the cooling of the Earth's BL surface and enhance IR cooling process (up to 30 W/m^2) because of dehydration at the higher levels by LSN (Girard & Blanchet, 2001a, 2001b).

The issues with light snow measurements (here called missing precipitation), and its effects on Earth's heat and energy budgets can be significant. The global mean precipitation results based on spatial and temporal distribution analysis, using satellite observations and gauge measurements over the 25 year time period 1979 to 2004, is found to be about 2.61 mm/day (from Global Precipitation Climatology Project (GPCP) data, Gruber & Levizzani, 2008). This PR suggests that missing LSN rate, dependent on density and shape of ice crystals, can affects PR uncertainty more than 50%, if the mean PR is <1 mm/day over Arctic regions (Walsh, Kattsov, Chapman, Govorkova, & Pavlova, 2002). The results suggest that snow measurements need to be improved down to 0.01 mm/h.

5.6 Overview of Snow Observations and Numerical Models

This section will emphasize issues related to light snow/snow predictions from forecast models and climate models, and also to solid precipitation measurements.

5.6.1 Numerical Weather Prediction Models

Precipitation prediction from Numerical Weather Prediction (NWP) models is strongly influenced by choice of cloud microphysical (CMP) schemes that usually predict ice mixing ratio and ice crystal number concentrations based on an assumed size distribution and a mass-length relationship. These CMP schemes are related to prescribed-values of physical parameters for various particle shapes (Gultepe, Axisa, Heymsfield, & McFarquhar, 2017). Precipitation formation is usually defined by autoconversion algorithms (Liu, Daum, & McGraw, 2006) using a prescribed ice crystal size of 200 μm (Gultepe et al., 2017). Microphysical schemes representing bulk and detailed microphysical parameterizations are based on one-moment, two-moment, and three-moment approaches applied to assumed size distributions and prognostic equations for one or more cloud microphysical parameters. Lately, new parameterizations (Harrington, Sulia, & Morrison, 2013a, 2013b; Milbrandt and Yau 2015; Morrison and Milbrandt, 2015; Morrison et al., 2015; Sulia, Morrison, & Harrington, 2015) based on single particle microphysics (called predicted particle properties (P3) scheme), specifically habit dependent, are being developed for various particle shapes in a bulk microphysical scheme. The microphysical and dynamical evolution of the cloud, simulated using the adaptive habit scheme, is compared with results based on spherical particle assumption with a bulk ice density or a reduced density, and mass—dimensional parameterization (Sulia et al., 2015). Their work showed that the adaptive habit method returns an increased (decreased) ice (liquid) mass as compared to traditional schemes. All these new parameterizations are not available for weather forecast operations but promising to improve ice cloud microphysics and eventually light snow precipitation.

Here, using prescribed microphysical relationships for ice crystal shapes and generalized gamma size distributions, the issues related to snow amount measurements and simulations are discussed. To emphasize how sensitive are the forecast models for solid precipitation prediction, the work of McMillen and Steenburgh (2015) based on National Center for Atmospheric Research (NCAR) Weather Research and Forecasting (WRF) simulations is used. Simulation of the Great Salt Lake snow storm that occurred on October 27, 2010 and an ice fog event with very light snow conditions studied by Gultepe et al. (2015) are summarized to indicate how difficult it is to predict the LSN amount. The later showed that cloud IWC and N_i, q_v, and related visibility parameters based on various CMP schemes (e.g., Milbrandt & Yau, 2005; Thompson, Field, Rasmussen, & Hall, 2008; Thompson, Rasmussen, & Manning, 2004, and Morrison et al., 2005; Morrison, Thompson, & Tatarskii, 2009) were at least 50% different each other. The work of McMillen and Steenburgh (2015) utilizing the NCAR WRF model based on Thompson et al. (2008) scheme (THOM), WRF with Double-Moment 6-class MP scheme (WDM6; Lim & Hong, 2010) produced larger snow amount compared to gauge observations. Overall, areal average precipitation coverage for PA >10 mm/day was overpredicted by the schemes of MORR (Morrison et al., 2009), GODD (Tao & Simpson, 1989; Goddard Scheme), and WDM6 compared to the THOM scheme (Figs. 23 and 24) based on various threshold values of PA.

Based on study of McMillen and Steenburgh (2015), radar-estimated precipitation (REP) from the lowest elevation angle reflectivity from the Promontary Point Radar and gauge based snow amounts are shown in Fig. 22 for the period of 02:30–17:00 UTC on October 27, 2010. In this figure, gauge measurements shown with unfilled circles (white ones) show no snow, indicated by <2 mm SWE. Most of the unfilled circles were covered by REP values >2 mm SWE amount. Large differences indicate undercatch of snow amount because of wind effects but estimation of SWE by the radar technique was also influenced by the fixed relationship between PR and Z, and other errors related to beam blockage, evaporation, ground-clutter contamination etc. The McMillen and Steenburgh (2015) study clearly showed that total condensate of hydrometeors and snow

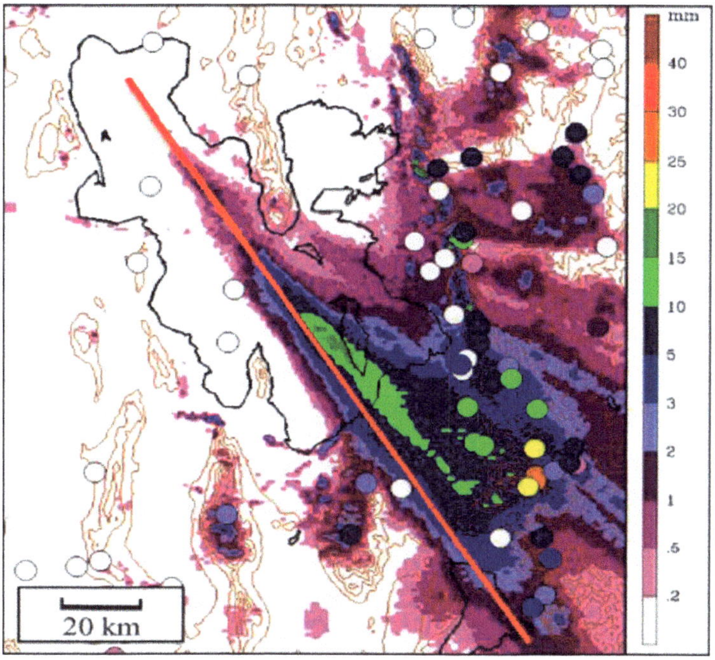

Figure 22 The radar estimated precipitation (mm) during snow storm period of 0230–17:00 UTC (coordinated universal time) on October 27, 2010 with gauge-measured precipitation indicated by *color-filled circles*. *Red line* overlaid as reference. *It is adapted from McMillen, J. D., & Steenburgh, W. J. (2015). Impact of microphysics parameterizations on simulations of the 27 October 2010 Great Salt Lake–effect snowstorm.* Weather Forecasting, 30, *136–152 by permission of AMS, Boston, MA.*

only profiles from all CMP schemes resulted in significant variability (Fig. 23). Both total condensate and snow amounts averaged over the time period had at least a factor of two differences among various CMP schemes. Fig. 24 compares the gauge SWE amounts overlaid on WRF-based CMP simulation results. In this figure, gauge based SWE with PA <0.2 mm does not match with well WRF simulations (unfilled circles seen left bottom). When WRF simulations indicate light precipitation, gauges SWE amount indicates almost 2 times more SWE amount (right top in all boxes in Fig. 24) compared to gauges SWE amount.

Overall, the present work suggests that based on biases related to gauge measurements and issues related to microphysical algorithms used in the NWPs, LSN prediction using NWPs needs more research and improvement of both observations and CMP schemes.

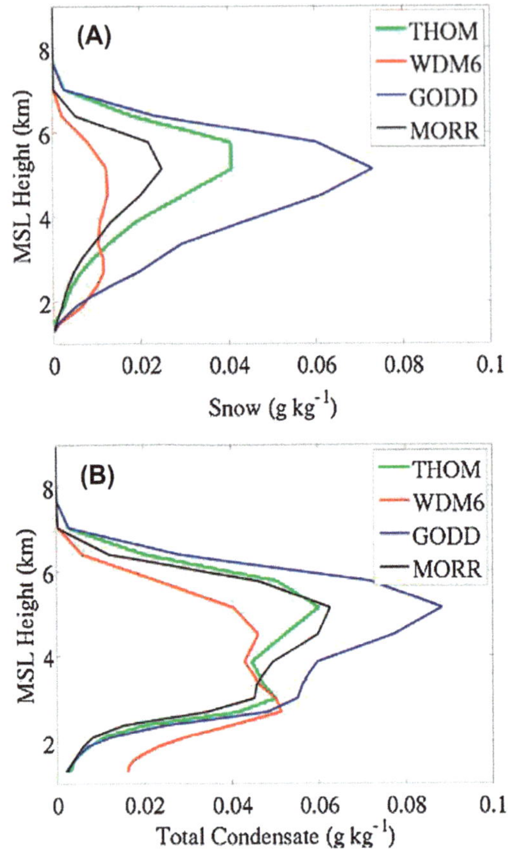

Figure 23 Vertical profiles of subdomain and storm period–averaged mixing ratio of (A) snow and (B) total condensate (cloud ice + cloud liquid water + rain + snow + graupel). *It is adapted from McMillen, J. D., & Steenburgh, W. J. (2015). Impact of microphysics parameterizations on simulations of the 27 October 2010 Great Salt Lake–effect snowstorm.* Weather Forecasting, 30, *136–152 by permission of AMS, Boston, MA.*

5.6.2 Climate Models

The light snow observations cannot be obtained accurately because of limitations in measuring snow amounts less than 0.5 mm/hand predicted using climate models because of uncertainties in precipitation parameterization. In fact, many observations provided in Arctic regions represent 12 h or daily accumulated snow amounts (Bogdanova et al., 2002). Serreze and Hurst (2000) studied issues related to Arctic precipitation and its prediction. They stated clearly that a major problem in understanding variability in

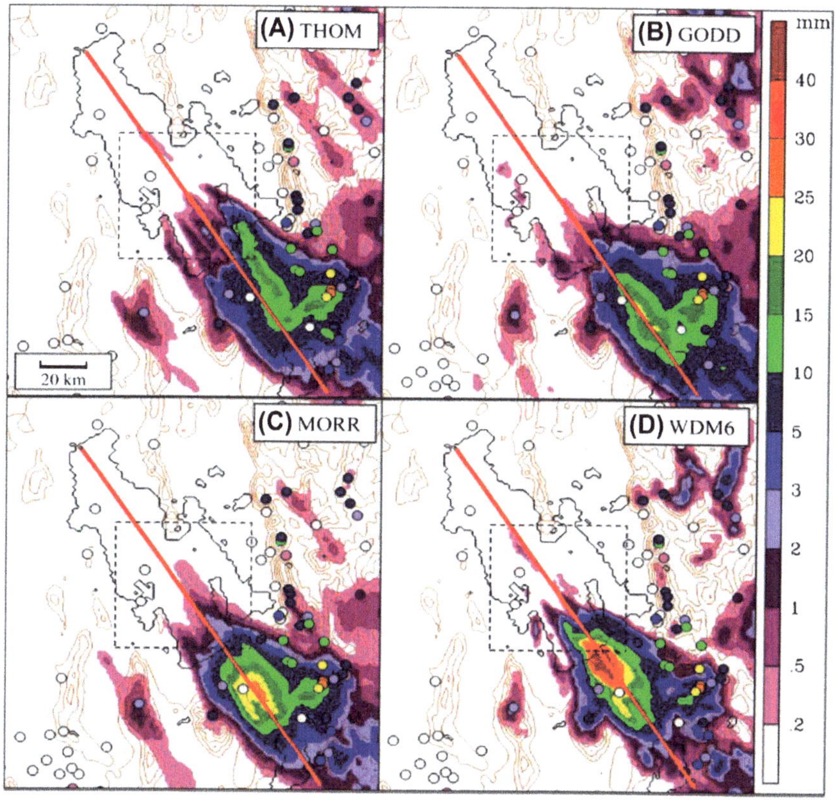

Figure 24 Simulated precipitation (mm) during the snow storm period from (A) THOM, (B) GODD, (C) MORR, and (D) WDM6 microphysical schemes. Gauge-measured precipitation indicated by *color-filled circles*. *Red line* and subdomain outline overlaid for reference. *It is adapted from McMillen, J. D., & Steenburgh, W. J. (2015). Impact of microphysics parameterizations on simulations of the 27 October 2010 Great Salt Lake–effect snowstorm.* Weather Forecasting, 30, *136–152 by permission of AMS, Boston, MA.*

the Arctic hydrologic budget is related to the uncertainty in precipitation estimation, and especially in winter precipitation. Weighing gauges or tipping buckets have limitations related to obtaining snow measurements, including undercatch of solid precipitation because of wind. Although it is possible to correct solid precipitation measurements for the effects of winds, gauge types, and wetting losses (Groisman, Koknaeva, Belokrylova, & Karl, 1991) based on empirical relationships, the density of observing sites over the Arctic regions is the source of one of the major problems. Because of the low station density, Arctic areas are usually inadequate for accurate mean snow amount calculations (Serreze & Hurst, 2000) and for model simulation validations.

Snow measurements from instruments, as provided above, can include large uncertainty that includes wind-induced undercatch, wetting loss, and missing precipitation amount based on lower threshold values from sensors (Bogdanova et al., 2002; Gultepe et al., 2015). Usually when trace amount precipitation occurs, it is assumed to be zero because sensors cannot detect them or a fixed value of precipitation is added to the total snow amount. Wind corrections are restricted to horizontal wind speeds of less than 6.5 m/s (Bogdanova et al., 2002). Blowing snow and blizzard are main sources of increased precipitation that result in overestimation of precipitation by the gauges (Bogdanova et al., 2002; Gultepe et al., 2015; Yang et al., 1998; Yang, Goodison, et al., 1999; Yang, Ishida, et al., 1999). Bogdanova et al. (2002) and Golubev and Bogdanova (1996) provided a set of equations for bias correction for solid precipitation, but these equations are mainly empirical relationships that are derived for certain types of sensors nd assumptions. The Bogdanova et al. (2002) study provides the corrections for wetting, evaporation, condensation, and trace precipitation as a function of RH but these cannot be hold because RH can be much lower when light snow occurs. Their results are also given over 12-h time periods but accumulation amount of light snow can be much larger if shorter time scales are considered. The corrections based on their work can be up to 12% of measured precipitation when the above issues are considered. False precipitation can be about 30% of annual precipitation when blowing snow and blizzards are considered. The Bogdanova et al. (2002) results did not show result of missing precipitation because of light snow conditions (<0.5 mm/h) that could not be measured accurately with weighing gauges.

The Atmospheric Model Intercomparison Project (AMIP) models showed overestimation of annual mean Arctic precipitation as compared to observations (Fig. 25). Walsh, Kattsov, Portis, and Meleshko (1998) and Walsh, Vavrus, and Chapman (2005), who used AMIP model simulations, suggested that monthly precipitation simulated for the Arctic Ocean differs by a factor of two among the models. For subregions, these differences among the models were much higher. The observations of precipitation data used in AMIP validations were corrected for wind, evaporation, wetting, and trace amounts of precipitation over 0.5ox0.5° grid areas (Legates & Willmott, 1990). Khrol (1996) also provided observational data with cold season corrections applied, and the values of their work were found to be larger than that of Legates and Willmott (1990). The monthly mean observed values and

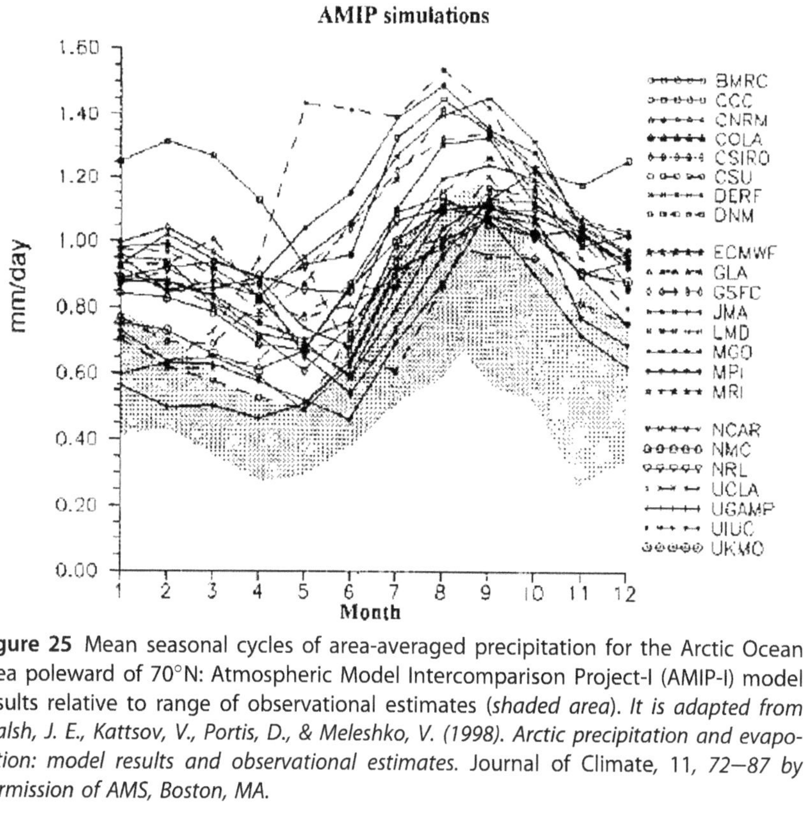

Figure 25 Mean seasonal cycles of area-averaged precipitation for the Arctic Ocean area poleward of 70°N: Atmospheric Model Intercomparison Project-I (AMIP-I) model results relative to range of observational estimates (*shaded area*). *It is adapted from Walsh, J. E., Kattsov, V., Portis, D., & Meleshko, V. (1998). Arctic precipitation and evaporation: model results and observational estimates.* Journal of Climate, 11, *72–87 by permission of AMS, Boston, MA.*

averages over the areas were also provided by Jaeger (1983, pp. 129–140) and Vowinckel and Orvig (1970), respectively. Walsh et al. (1998) using their data from 1979 to 1988 time period suggested that AMIP model results are 50–100% greater than observed estimates of Jaeger (1983, pp. 129–140) and Vowinckel and Orvig (1970) in the 50–70°N latitudes. Although observations are mainly corrected for biases associated with wind and other errors (Yang & Ohata, 2001), models usually generate very large values of snow amounts compared to observations. These comparisons did not address trace amounts of solid precipitation but total snow amounts.

Based on measurement issues related to solid precipitation when PR < 0.5 mm/h, it is very difficult to validate model-based predictions. In fact, biases in observations of snow measurements can be much larger than the LSN values. Walsh et al. (2002) used 13 global atmosphere-only climate models (AMIP-II) and eight coupled atmosphere-ocean-ice models from the data distribution center of the Intergovernmental Panel on Climate

Change (IPCC). Their results suggested that T variances in coupled models were larger than T of the uncoupled models. Precipitation amount is still overestimated compared to AMIP-I results and the bias was larger in the coupled models. Walsh et al. (2002) stated that major issues for assessment of simulated precipitation in high latitudes are related to uncertainty in snow measurement, and these are related to solid precipitation under the windy conditions and blowing snow effects. Their work citing Bogdanova et al. (2002) stated that precipitation measurements can be in error by up to 50% uncertainty. Based on ice fog studies, Gultepe et al. (2015) suggested that light snow particles may have fall speeds of 10 cm/s and this suggests that LSN amount may not be measured accurately even under wind speeds of less than 1–2 m/s. Fig. 26 shows that the AMIP-II models estimate the precipitation amount to be much larger as compared to measured snow amounts, even if corrections are applied to Arctic snow measurements (Bryazgin, 1976; Khrol, 1996) above 65°N.

Large-scale climate models have severe issues regarding predicting T and cloud fraction (c_f) over the Arctic regions. Tao, Walsh, and Chapman (1996) and Tao et al. (2003) using Arctic seasonal T and c_f observations based on 10-year (1979–88) data set and simulation results from the 19 AMIP models for winter, i.e., during the time period of December, January, and February (DJF) found that differences in T and c_f were about more than 20°C and 50%, respectively. Fig. 27 shows T (A) and c_f (B) plotted versus latitude and months, respectively, for 10 climate models. The heavy solid line is for observed values. Cold bias during winter at about 1–3°C exists (Tao et al., 1996) over the Arctic Ocean and a maximum at 3.3°C (sd = 3.4°C) over northern Eurasia. Clearly, large-scale precipitation parameterization (Sundqvist, Berge, & Kristjánsson, 1989) is strongly related to both c_f (also RH) and T and these parameters have very large biases.

The precipitation prediction from climate models uses a precipitation parameterization based on Sundqvist et al. (1989) that is a function of c_f, LWC, RH, and T. Any bias related to these parameters affects winter precipitation. It is shown in the work of Vavrus and Waliser (2008) that climate models (including National Center for Atmospheric Research (NCAR) Community Atmosphere Model Version-3 (CAM3)) generate excessive c_f in the BL low cloud amount. They modified the traditional use of the $c_f = f(RH, RH_{min})$ relationship where RH_{min} is an assumed threshold value of RH of 0.8 over land and 0.9 in other places. Based on Thompson and Pollard (1995) method, a threshold value of specific humidity (q_{ct}) of 0.003 kg/kg (3 g/kg) is used to improve c_f for low clouds. Performing this

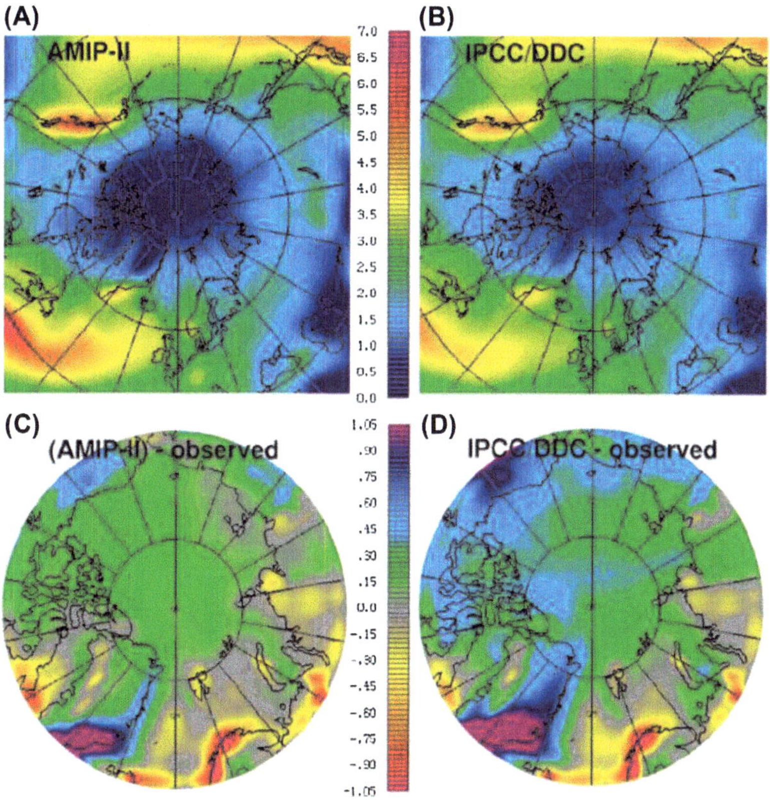

Figure 26 Composite annual mean precipitation rate (mm/day) from the uncoupled Atmospheric Model Intercomparison Project-II (AMIP-II) models (A) and the coupled IPCC/DDC models. (B). Corresponding departures from observational climatology of Bryazgin (1976) are shown in (C) and (D). *It is adapted from Walsh, J. E., Kattsov, V. M., Chapman, W. L., Govorkova, V., & Pavlova, T. (2002). Comparison of Arctic climate simulations by uncoupled and coupled global models.* Journal of Climate, 15, *1429–1446 by permission of AMS, Boston, MA.*

new calculation, bias in c_f is reduced up to 85%. Based on control runs of the Community Climate System Model Version-3 (CCSM3; Vavrus & Waliser, 2008), low cloud amount bias is predicted more than 30% in northern latitudes, and that amount resulted in T bias at about 2–10°C. These results clearly showed that because of high biases in both T and cloud cover, LSN amount cannot be accurately predicted by using climate models. The new parameterization based on an assumed q_c reduced the bias by up to 50% during DJF (December, January, and February) time period. This uncertainty in fact exists in many climate models as described by Vavrus and Waliser

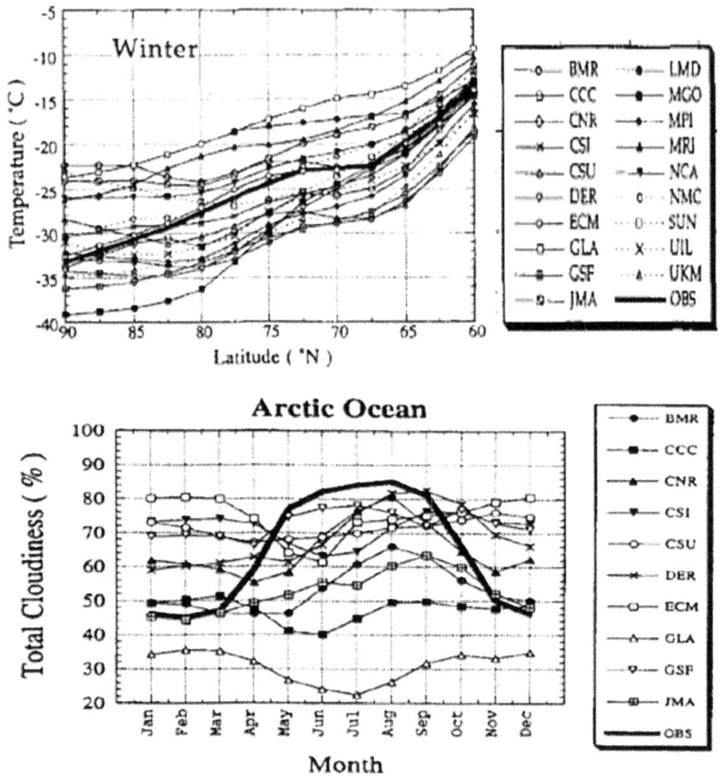

Figure 27 Zonal average of the 10-year seasonal mean temperatures of the Atmospheric Model Intercomparison Project (AMIP) models for winter (Dec, Jan, and Feb) (top box). The legends shown on the right are for models used in the work. Bottom box is for annual cycle of monthly mean cloud total cloudiness (%) over the Arctic Ocean versus months. *Heavy solid line* is for observed values of cloud cover for 75°N-85°N of Vowinckel and Orvig (1970). *It is adapted from Tao, X., Walsh, J. E., & Chapman, W. L. (1996). An assessment of global climate model simulations of Arctic air temperatures.* Journal of Climate, 9, *1060–1076 by permission of AMS, Boston, MA.*

(2008). A latest work on cloud cover predicted by the climate models and observations also showed that cloud cover ranges from 40% to 90% for a given month (Christensen et al., 2016), suggesting that winter precipitation from models still needs to be improved before it can be used for LSN predictions.

The bias in precipitation amount was decreased with the new parameterization using a threshold for q_c but snow precipitation simulations and measurements are still a major issue for cold climatic regions, because the precipitation amount from a model grid area is related to cloud cover amount and thermodynamic parameters. Prediction of LSN amount from a climate model is still an important issue because of very large biases related to T, q_c, and c_f; therefore, climate simulations were not performed in this study. Note that cloud cover parameterizations do not provide information on optical depth, water amount or cloud thickness, and, therefore, large uncertainties can exist when LSN conditions are considered.

6. CONCLUSIONS

Precipitation processes are important for climate change assessment, hydrological cycle representing water cycle in the atmosphere and over the Earth surface, NWP model validations, and aviation applications. The LSN (defined as PR < 0.5 mm/h) precipitation in cold climates usually cannot be measured accurately because of instrumental limitations. These limitations include lack of sensor calibrations for cold weather conditions, and unreliable response of the optical sensors to the cold and harsh environments.

The missing part of the total snow precipitation amount, including ice crystals from clouds, ice fog crystals, and diamond dust particles, is important for hydrological assessment and weather applications. The LSN precipitation can also be responsible for discrepancies in precipitation retrievals between remote sensing platforms and in situ observations, as well as between model-based predictions and in situ-based observational analysis results.

Other issues such as time and space scale variability of cloud parameters, the chaotic nature of snow formation processes, measurement errors, as well as processing the data with interpolation and areal sampling errors (Rasmussen et al., 1982; Rudolf et al., 1994) can affect the snow precipitation trends and its energy equivalent values for studying global precipitation anomalies accurately. It is possible that future precipitation studies using Arctic observing satellite measurements (Trishchenko, Garand, & Trichtchenko, 2011) can improve the light snow observations but limitations in retrieval techniques related to particle spectral physics and ice cloud processes can still affect the quality of the LSN quantification.

Here, the LSN events that occurred during the field project for two cases are provided. Then, a summary of entire snow events for the winter of 2013–14 (Fig. 21) is given. Overall it is found that LSN events can be better represented by optical spectral precipitation probes (OSPP) and PWOS. The following conclusions based on the results can be obtained from the current work:

1. Light snow precipitation measurements over cold climates cannot be obtained accurately using weighing gauges because of collection efficiency, sensor sensitivity, environmental factors such wind-induced errors, and calibration issues.
2. Present weather optical sensors such as PWD and SWS can be used to detect and measure LSN precipitation amounts <0.5 mm/h. Further validations and research can improve the outcome of this work although results are consistent those from Zhang et al. (2015).
3. The impact of light snow measurements (Fig. 21) on the assessment of climate change needs to be accurately addressed. This impact can be responsible for Arctic cooling processes and modifying frontogenesis events.
4. LSN can significantly affect visibility conditions over the Arctic regions because of large amount of small ice crystals.
5. Integration of precipitation observations over various size ranges obtained from various instruments is recommended to better predict and characterize light precipitation conditions, visibility, and model performance assessments.
6. Measurements from the instruments such as GCIP can improve detecting and measuring ice crystals and ice fog, as well as light snow precipitation spectral characteristics. Therefore, these observations can serve to improve assessment of LSN effect on weather and climate issues.
7. The 3D wind conditions in the analysis of light precipitation should be considered because of the way small particles move with wind flow. This effect is minimized when optical sensors e.g., OPWS sensors are used for light precipitation measurements. For the GCIP case, this condition can be ignored because of air flow into inlet at 25 m/s.
8. Missing precipitation due to measurement limitations of ice crystals and drizzle can significantly affect the mean precipitation values and precipitation duration over the colder and continental climatic regions. These limitations cannot be simulated by numerical models (Gultepe et al., 2015) using current model physical algorithms.
9. Aviation applications related to visibility may be impacted by large uncertainty in snow measurements during which LSN can affect the visibility significantly. Visibility-PR relationships for heavy snow conditions cannot be valid for LSN conditions.

In the future, observations from various projects will be used for more detailed quantitative assessments of the LSN effects on weather and climate issues over the Arctic regions.

APPENDIX

Table A1 Nomenclature

Acronyms	**Definition**
A	Constant as 2.49
a	Constant as 0.3407 in Weibull distribution
A_c	$3.6 \times 10^{+4}$ constant in precipitation rate calculation
B	Constant as 4.68×10^4
b	0.67 in Weibull distribution
CAPE	Convectively available potential energy
c_f	Cloud fraction
D	Diameter of spherical ice crystals
D_o	Median diameter
DAS	Data acquisition system
DELSN	Direct effect of light snow
DMT	Droplet measurements technologies
DSP	Digital signal processor
E	Polar regions based on Köppen classification
EC	Environment Canada
ET	Polar tundra based on Köppen classification
EF	Polar ice cap based on Köppen classification
EM	Polar marine based on Köppen classification
FD12p	Vaisala present weather sensor
FMD	Fog measuring device
FRAM	Fog remote sensing and modeling project
GCIP	Ground cloud imaging probe
GPCP	Global precipitation climatology project
HSN	Heavy snow
ELSN	Indirect effect of light snow
ILW	Integrated liquid water
IN	Ice nuclei
IPCC	International panel climate change
IR	Infrared radiation
IWC	Ice water content
IWV	Integrated water vapor

(Continued)

Table A1 Nomenclature—cont'd

Acronyms	Definition
K_c	Conversion constant as 1/86,400 day
L_{ice}	Latent heat of sublimation of ice crystals
LPM	Laser precipitation monitor
LSN	Light snow
LWP	Liquid water path
LWC	Liquid water content
m(D)	Ice crystal mass at diameter D
m	Ice crystal or snow mass
MWR	Microwave radiometer
N(D)	Particle spectra
NFL	NewFoundLand
N_i	Ice crystal number concentration
N_{ti}	Total ice crystal number concentration
NN	Neural network method
NWP	Numerical weather prediction
NWS	National weather service
PA	Accumulated precipitation
PADS	Particle analyzing and display software
PMWR	Profiling microwave radiometer
PR	Precipitation rate
PR_H	PR for heavy snow
PR_L	PR for light snow
PR_{sn}	PR for snow
PR_T	PR for total snow PR
PSD	Particle size distribution
PWD	Present weather sensor from Vaisala Inc.
q_c	Specific humidity
q_{ct}	Specific humidity threshold
REP	Radar-estimated precipitation
RH_w	Relative humidity with respect to water
RH_i	Relative humidity with respect to ice
RH_{min}	Assumed value for RH used in cloud fraction estimation
SAAWSO	Satellite application for Arctic weather and SAR operations
SAR	Search and rescue
OPWS	Optical present weather sensor
OSPP	Optical spectral precipitation probes
SWE	Snow water equivalent
SWS	Metek present weather sensor
T	Temperature
Td	Dewpoint temperature
T_{ir}	Infrared cloud base temperature
T_{amb}	Ambient air temperature

Table A1 Nomenclature—cont'd

Acronyms	Definition
TAS	True air speed
V_f	Particle fall velocity
V_i	Ice crystal volume
Vis	Visibility
x	Variable in the Weibull precipitation distribution
Q_e	Equivalent heat release because of snow melting
P_s	Density of snow

ACKNOWLEDGMENTS

Funding for this work was provided by the Canadian National Search and Rescue Secretariat under the SAR project # SN 201221/EC S2001-T104-H029-MR029-165 and Environment Canada.

REFERENCES

Aaltonen, A., Elomaa, E., Tuominen, A., & Valkovuori, P. (1993). Measurement of precipitation. In B. Sevruk, & M. Lapin (Eds.), *Proceedings of the symposium on precipitation and evaporation* (pp. 42–46). Bratislava, Slovakia: Slovak Hydrometeorlogical Institute and Swiss Federal Institute of Technology.

Alter, J. C. (1937). Shielded storage precipitation gauges. *Monthly Weather Review, 65*, 262–265.

Ballinger, T. J., Schmidlin, T. W., & Steinhoff, D. F. (2013). The polar marine climate revisited. *Journal of Climate, 26*, 3935–3952.

Barry, R. G. (1981). *Mountain weather and climate.* London: Methuen & Corp. Ltd, 313 pp.

Bengtsson, L., Hagemann, S., & Hodges, K. I. (2004). Can climate trends be calculated from reanalysis data? *Journal of Geophysical Research, 109*, D11111. http://dx.doi.org/10.1029/2004JD004536.

Bengtsson, L., Hodges, K. I., & Hagemann, S. (2004). Sensitivity of the ERA-40 reanalysis to the observing system: determination of the global atmospheric circulation from reduced observations. *Tellus, 56A*, 456–471.

Bogdanova, E. G., Ilyin, B. M., & Dragomilova, I. V. (2002). Application of a comprehensive bias-correction model to precipitation measured at Russian north pole drifting stations. *Journal of Hydrometeorology, 3*, 700–713.

Boudala, F. S., Rasmussen, R., Isaac, G. A., & Scott, B. (2014). Performance of hot plate for measuring solid precipitation in complex terrain during the 2010 Vancouver Winter Olympics. *Journal of Atmospheric and Oceanic Technology, 31*, 437–446.

Brandes, E. A., Ikeda, K., Zhang, G., Schönhuber, M., & Rasmussen, R. M. (2007). A statistical and physical description of hydrometeor distributions in Colorado snowstorms using a video disdrometer. *Journal of Applied Meteorology and Climatology, 46*, 634–650.

Brawn, D., & Upton, G. (2008). Estimation of an atmospheric gamma drop size distribution using disdrometer data. *Atmospheric Research, 87*(1), 66–79.

Bryazgin, N. N. (1976). Yearly mean precipitation in the Arctic region accounting for measurement errors. *Proceedings of the Arctic and Antarctic Research Institute, 323*, 40–74.

Christensen, M. W., Behrangi, A., L'Ecuyer, T., Wood, N. B., Lebsock, M. D., & Stephens, G. L. (2016). Arctic observation and reanalysis integrated system: a new data

product for validation and climate study. *Bulletin of the American Meteorological Society,* 907–915.
Cimini, D., Campos, E., Ware, R., Albers, S., Graziano, G., Oreamuno, J., ... Westwater, E. (2011). Thermodynamic atmospheric profiling during the 2010 Winter Olympics using ground-based microwave radiometry. *IEEE Transaction on Geosciences and Remote Sensing, 49,* 4959–4969.
Cimini, D., Nelson, M., Güldner, J., & Ware, R. (2015). Forecast indices from ground-based microwave radiometer for operational meteorology. *Atmospheric Measurement Techniques, 8,* 315–333.
Cimini, D., Westwater, E., & Gasiewski, A. (2010). Temperature and humidity profiling in the Arctic using ground-based millimeter-wave radiometry and 1DVAR. *IEEE Transaction on Geosciences and Remote Sensing, 48,* 1381–1388.
DMT Inc. (2004). *Manual for clouid imaging probe.* Boulder, CO, US: DMT Inc. DOC-0028 REV D. 43 pp.
Girard, E., & Blanchet, J. P. (2001a). Microphysical parameterization of Arctic diamond dust, ice fog, and thin stratus for climate models. *Journal of the Atmospheric Sciences, 58,* 1181–1198. Agriculture applications.
Girard, E., & Blanchet, J. P. (2001b). Simulation of Arctic diamond dust, ice fog, and thin stratus using an explicit aerosol–cloud–radiation model. *Journal of the Atmospheric Sciences, 58,* 1199–1221.
Golubev, V. S., & Bogdanova, E. G. (1996). Point measurements of solid precipitation. In *Proc. workshop on the ACSYS solid precipitation climatology project* (pp. 18–29). Reston, VA: World Climate Research Programme.
Goodison, B. E. (1981). Compatibility of Canadian snowfall and snowcover data. *Water Resources Research, 17,* 893–900.
Goodison, B. E., Louie, P. Y. T., & Yang, D. (1998). *WMO solid precipitation measurement intercomparison, final report WMO instruments and observing methods.* Rep. 67, WMO/TD-872, 212 pp.
Goodison, B. E., & Yang, D. (1995). In situ measurement of solid precipitation in high latitudes: the need for correction. In *Proceedings of ACSYS solid precipitation climatology project workshop, Reston, VA* (pp. 1–12). Geneva: WMO. WMO/TD No. 739.
Groisman, P. Y., Koknaeva, V. V., Belokrylova, T. A., & Karl, T. R. (1991). Overcoming biases of precipitation measurement: a history of the USSR experience. *Bulletin of the American Meteorological Society, 72,* 1725.
Gruber, A., & Levizzani, V. (2008). *Assessment of global precipitation products. A project of the world climate research programme. Global energy and water cycle experiment (GEWEX) radiation panel cooperative institute for climate studies.* College Park, Maryland: Earth System Science Interdisciplinary Center, University of Maryland. WCRP-128 WMO/TD-No. 1430. 50 pp.
Güldner, J., & Spänkuch, D. (2001). Remote sensing of the thermodynamic state of the atmospheric boundary layer by ground-based microwave radiometry. *Journal of the Atmospheric and Oceanic Technology, 18,* 925–933.
Gultepe, I. (2008). Measurements of light rain, drizzle and heavy fog. In S. Michaelides (Ed.), *A chapter in precipitation: Advances in measurement, estimation and prediction.* Springer-Verlag, ISBN 978-3-540-77654-3.
Gultepe, D., Axisa, A. J., Heymsfield, G., & McFarquhar, F. (2017). Ice phase precipitation. In *Chapter 6: Ice formation and evolution in clouds and precipitation: Measurement and modelling challenges.* Boston: AMS Meteorological Monographs, AMS (Accepted).
Gultepe, I., Isaac, G. A., Joe, P., Kucera, P., Thériault, J., & Fisico, T. (2014b). Roundhouse (RND) mountain top research site: measurements and uncertainties for winter Alpine weather conditions. *Pure and Applied Geophysics, 171,* 59–85. http://dx.doi.org/10.1007/s00024-012-0582-5.

Gultepe, I., Kuhn, T., Pavolonis, M., Calvert, C., Gurka, J., Isaac, G. A., … Bernstein, B. (2014a). Ice fog in Arctic during FRAM-IF project: aviation and nowcasting applications. *Bulletin of the American Meteorological Society, 95*, 211–226.

Gultepe, I., Pearson, G., Milbrandt, J. A., Hansen, B., Platnick, S., Taylor, P., … Cober, S. G. (2009). The fog remote sensing and modeling (FRAM) field project. *Bulletin of the American Meteorological Society, 90*, 341–359.

Gultepe, I., Tardif, R., Michaelides, S. C., Cermak, J., Bott, A., Bendix, J., … Cober, S. G. (2007). Fog research: a review of past achievements and future perspectives. *Journal of Pure and Applied Geophysics, 164*, 1121–1159.

Gultepe, I., Zhou, B., Milbrandt, J., Bott, A., Li, Y., Heymsfield, A. J., … Cermak, J. (2015). A review on ice fog: observations and modeling. *Atmospheric Research*, 2–19. http://dx.doi.org/10.1016/j.atmosres.2014.04.014.

Harrington, J. Y., Sulia, K., & Morrison, H. (2013a). A method for adaptive habit prediction in bulk microphysical models. Part I: theoretical development. *Journal of the Atmospheric Sciences, 70*, 349–364.

Harrington, J. Y., Sulia, K., & Morrison, H. (2013b). A method for adaptive habit prediction in bulk microphysical models. Part II: parcel model corroboration. *Journal of the Atmospheric Sciences, 70*, 365–376.

Heymsfield, A. J., & Platt, C. M. R. (1984). A parameterization of the particle size spectrum of ice clouds in terms of the ambient temperature and the ice water content. *Journal of the Atmospheric Sciences, 41*, 846–855.

Heymsfield, A. J., Zadelhoff, G. V., Donovan, D. P., Fabry, F., Hogan, R. J., & Illingworth, A. J. (2007). Refinements to ice particle mass dimensional and terminal velocity relationships for ice clouds. Part II: evaluation and parameterizations of ensemble ice particle sedimentation velocities. *Journal of the Atmospheric Sciences, 64*, 1068–1088.

Huffman, G. J., Adler, R. F., Rudolf, B., Schneider, U., & Keehn, P. R. (1995). Global precipitation estimates based on a technique for combining satellite-based estimates, rain gauge analysis, and NWP model precipitation information. *Journal of Climate, 8*, 1285–1295.

Intrieri, J. M., & Shupe, M. D. (2004). Characteristics and radiative effects of diamond dust over the Western Arctic Ocean region. *Journal of Climate, 17*, 2953–2960.

IPCC. (2007). Climate change 2007: the physical science basis. In *Contribution of Working Group I to the Fourth Assessment Report of the Intergovernmental Panel on Climate Change*. Cambridge University Press, 996 pp.

Jaeger, L. (1983). Monthly and areal patterns of mean global precipitation. In A. Street-Perrot, M. Beran, & R. Ratcliffe (Eds.), *Variations in the Global Water Budget*. D. Reidel.

Jaffrain, J., & Berne, A. (2011). Experimental quantification of the sampling uncertainty associated with measurements from PARSIVEL disdrometers. *Journal of Hydrometeorology, 12*, 352–370.

Kajikawa, K., & Heymsfield, A. J. (1989). Aggregation of ice crystals in cirrus. *Journal of the Atmospheric Sciences, 46*, 3108–3121.

Khrol, V. P. (1996). *Atlas of water balance of the northern polar area*. Gidrometeoizdat, 81 pp.

Knupp, K., Ware, R., Cimini, D., Vandenberghe, F., Vivekanandan, J., Westwater, E., & Coleman, T. (2009). Ground-based passive microwave profiling during dynamic weather conditions. *Journal of the Atmospheric and Oceanic Technology, 26*, 1057–1073.

Legates, D. R., & Willmott, C. J. (1990). Mean seasonal and spatial variability in gage-corrected global precipitation. *International Journal of Climatology, 10*, 111–133.

Lim, K. S. S., & Hong, S. Y. (2010). Development of an effective double-moment cloud microphysics scheme with prognostic cloud condensation nuclei (CCN) for weather and climate models. *Monthly Weather Review, 138*, 1587–1612.

Liston, G. E., & Hiemstra, C. A. (2011). The changing cryosphere: Pan-Arctic snow trends (1979–2009). *Journal of Climate, 24*, 5691–5712.

Liu, Y., Daum, P. H., & McGraw, R. (2006). Parameterization of the autoconversion process. Part II: generalization of sundqvisibilityt-type parametrizations. *Journal of the Atmospheric Sciences, 63*, 1103–1109.
Löhnert, U., Kneifel, S., Battaglia, A., Hagen, M., Hirsch, L., & Crewell, S. (2011). A multisensor approach toward a better understanding of snowfall microphysics: the TOSCA project. *Bulletin of the American Meteorological Society, 92*, 613–628.
LPM Technical Manual. (2007). *Thies Clima laser precipitation monitor manual*, 5.4110.xx.x00 V2.2x STD. 58 pp.
McMillen, J. D., & Steenburgh, W. J. (2015). Impact of microphysics parameterizations on simulations of the 27 October 2010 Great Salt Lake–effect snowstorm. *Weather Forecasting, 30*, 136–152.
Milbrandt, J. A., & Yau, M. K. (2005). A multimoment bulk microphysics parameterization. Part II: a proposed three-moment closure and scheme description. *Journal of the Atmospheric Sciences, 62*, 3065–3081.
Morrison, H., Curry, J. A., & Khvorostyanov, V. I. (2005). A new double-moment microphysics parameterization for application in cloud and climate models. Part I: description. *Journal of the Atmospheric Sciences, 62*, 1665–1677.
Morrison, H., & Milbrandt, J. A. (2015). Parameterization of ice microphysics based on the prediction of bulk particle properties. Part I: scheme description and idealized tests. *Journal of the Atmospheric Sciences, 72*, 287–311.
Morrison, H., Milbrandt, J. A., Bryan, G. H., Ikeda, K., Tessendorf, S. A., & Thompson, G. (2015). Parameterization of cloud microphysics based on the prediction of bulk ice particle properties. Part II: case study comparisons with observations and other schemes. *Journal of the Atmospheric Sciences, 72*, 312–339.
Morrison, H., Thompson, G., & Tatarskii, V. (2009). Impact of cloud microphysics on the development of trailing stratiform precipitation in a simulated squall line: comparison of one- and two-moment schemes. *Monthly Weather Review, 137*, 991–1007.
Politovich, M., Stankov, B., & Martner, B. (1995). Determination of liquid water altitudes using combined remote sensors. *Journal of Applied Meteorology, 34*, 2060–2075.
Randall, D., Krueger, S., Bretherton, C., Curry, J., Duynkerke, P., Moncrieff, M., … Wielicki, B. (2003). Confronting models with data: the GEWEX cloud systems study. *Bulletin of the American Meteorological Society, 84*, 455–469.
Rasmussen, R., Baker, B., Kochendorfer, J., Meyers, T., Landolt, S., Fischer, A. P., … Craig Sevruk, B. (1982). *Method of correction for systematic error in point precipitation measurement for operational use.* WMO No. 589. Geneva: WMO, 91 pp..
Rasmussen, R., Baker, B., Kochendorfer, J., Meyers, T., Landolt, S., Fischer, A. P., … Gutmann, E. (2012). How well are we measuring snow: the NOAA/FAA/NCAR winter precipitation test bed. *Bulletin of the American Meteorological Society, 93*, 811–829.
Rasmussen, R. M., Vivekanandan, J., Cole, J., Myers, B., & Masters, C. (1999). The estimation of snowfall rate using visibility. *Journal of Applied Meteorology, 38*, 1542–1563.
Roads, J. (February 2002). Closing the water cycle. *GEWEX Newsletter, 12*, 6–8.
Rohli, R. V., & Vega, A. J. (2011). *Climatology* (2nd ed.). Jones and Bartlett, 432 pp.
Rudolf, B., Hauschild, H., Rüth, W., & Schneider, U. (1994). Terrestrial precipitation analysis: operational method and required density of point measurements. In M. Desbois, & F. Desalmand (Eds.), *NATO ASI I/26, Global precipitations and climate change* (pp. 173–186). Berlin: Springer Verlag.
Rudolf, B., Hauschild, H., Rüth, W., & Schneider, U. (1996). Comparison of raingauge measurements, precipitation estimates from satellite data and model results. *Advances in Space Research, 18*, 53–62.
Rudolf, B., & Rubel, F. (2005a). Global precipitation. In M. Hantel (Ed.), *Landolt-Bornstein, numerical data and functional relationshipsObserved global climate* (pp. 11.1–11.53).

Berlin: Springer, ISBN 3-540-20206-4. extended online version (11.1—11.24, print version).

Rudolf, B., & Rubel, F. (2005b). Global precipitation. In , *Landolt-Bornstein-Group V Geophysics Series: Vol. 6. Chapter 11 in observed global climate* (pp. 1—24). Published by Springer Berlin Heidelberg, ISBN 978-3-540-20206-6.

Sanchez, J., Posada, R., Garcia-Ortega, E., Lopez, L., & Marcos, J. L. (2013). A method to improve the accuracy of continuous measuring of vertical profiles of temperature and water vapor density by means of ground-based microwave radiometer. *Atmospheric Research, 122*, 43—54.

Serke, D., Hall, E., Bognar, J., Jordan, A., Abdo, S., Baker, K., … Politovich, M. (2014). Supercooled liquid water content profiling case studies with a new vibrating wire sonde compared to a ground-based microwave radiometer. *Atmospheric Research, 149*, 77—87.

Serreze, M., & Hurst, C. (2000). Representation of mean Arctic precipitation from NCEP-NCAR and ERA reanalyses. *Journal of Climate, 13*, 182—201.

Sevruk, B. (1982). Methods of correction for systematic error in point precipitation measurement for operational use. *WMO Operational Hydrology Rep, 21*(Publ., 589), 91pp.

Sevruk, B., & Klemm, S. (1989). Types of standard precipitation gauges. In *Proc. Int. Workshop on Precipitation Measurement, WMO/Tech. Doc. 328* (pp. 227—236). St. Moritz, Switzerland, Department of Geography, Swiss Federal Institute of Technology, ETH Zurich.

Smith, C. D. (2009). The relationship between snowfall catch efficiency and wind speed for the Geonor T-200B precipitation gauge utilizing various wind shield configurations. In *Proc. 77th Western snow Conference, Canmore, AB, Canada* (pp. 115—121).

Solheim, F., Godwin, J., & Ware, R. (1998). Passive ground-based remote sensing of atmospheric temperature, water vapor, and cloud liquid water profiles by a frequency synthesized microwave radiometer. *Meteorologische Zeitschrift, 7*, 370—376.

Sterl, A. (2004). On the (in) homogeneity of reanalysis products. *Journal of Climate, 17*, 3866—3873.

Stoelinga, M. T., & Warner, T. T. (1999). Nonhydrostatic, meso-beta scale model simulations of cloud ceiling and visibility for an east coast winter precipitation event. *Journal of Applied Meteorology, 38*, 385—404.

Sulia, K. J., Morrison, H., & Harrington, J. Y. (2015). Dynamical and microphysical evolution during mixed-phase cloud glaciation simulated using the bulk adaptive habit prediction model. *Journal of the Atmospheric Sciences, 71*, 4158—4180.

Sundqvist, H., Berge, E., & Kristjansonn, J. E. (1989). Condensation and cloud parameterization studies with a mesoscale numerical weather prediction model. *Monthly Weather Review, 117*, 1641—1657.

Tao, W. K., et al. (2003). Microphysics, radiation and surface processes in the Goddard Cumulus Ensemble (GCE) model. *Meteorology and Atmospheric Physics, 82*, 97—137.

Tao, W. K., Simpson, J., & McCumber, M. (1989). An ice-water saturation adjustment. *Monthly Weather Review, 117*, 231—235.

Tao, X., Walsh, J. E., & Chapman, W. L. (1996). An assessment of global climate model simulations of Arctic air temperatures. *Journal of Climate, 9*, 1060—1076.

Tapiador, F. J., Turk, F. J., Petersen, W., Hou, A. Y., Ortega, E. G., Carlos, L. M., … Castro, M. (2012). Global precipitation measurement: methods, datasets and applications. *Atmospheric Research, 104—105*, 70—97.

Thompson, G. R., Field, P. R., Rasmussen, R. M., & Hall, W. D. (2008). Explicit forecasts of winter precipitation using an improved bulk microphysics scheme. Part II: Implementation of a new snow parameterization. *Monthly Weather Review, 136*, 5095—5114.

Thompson, S. L., & Pollard, D. (1995). A global climate model (GENESIS) with a land-surface (LSX). Part I: present climate simulation. *Journal of Climate, 8*, 732—761.

Thompson, G., Rasmussen, R. M., & Manning, K. (2004). Explicit forecasts of winter precipitation using an improved bulk microphysics scheme. Part I: description and sensitivity analysis. *Monthly Weather Review, 132*, 519—542.

Tian, Y., Peters-Lidard, C. D., & Eylander, J. B. (2010). Real-time bias reduction for satellite-based precipitation estimates. *Journal of Hydrometeorology, 11*, 1275–1285.

Trishchenko, A. P., Garand, L., & Trichtchenko, L. D. (2011). Three-apogee 16-h highly elliptical orbit as optimal choice for continuous meteorological imaging of polar regions. *Journal of Atmospheric and Oceanic Technology, 28*, 1407–1422.

Trouvilliez, A., Naaim-Bouvet, F., Bellot, H., Genthon, C., & Gallée, H. (2015). Evaluation of the FlowCapt acoustic sensor for the aeolian transport of snow. *Journal of Atmospheric and Oceanic Technology, 32*(9), 1630–1641.

Turner, D. (2007). Improved ground-based liquid water path retrievals using a combined infrared and microwave approach. *Journal of Geophysical Research, 112*, D15204. http://dx.doi.org/10.1029/2007JD008530.

Vavrus, S., & Waliser, D. (2008). An improved parameterization for simulating Arctic cloud amount in the CCSM3 climate model. *Journal of Climate, 21*, 5673–5687.

Vowinckel, E., & Orvig, S. (1970). The climate of the north polar basin. In S. Orvig (Ed.), *Climates of the polar regions: Vol. 14. World survey of climatology* (pp. 129–252). Elsevier.

Walsh, J. E., Kattsov, V. M., Chapman, W. L., Govorkova, V., & Pavlova, T. (2002). Comparison of Arctic climate simulations by uncoupled and coupled global models. *Journal of Climate, 15*, 1429–1446.

Walsh, J. E., Kattsov, V., Portis, D., & Meleshko, V. (1998). Arctic precipitation and evaporation: model results and observational estimates. *Journal of Climate, 11*, 72–87.

Walsh, J. E., Vavrus, S. J., & Chapman, W. L. (2005). Summary of a workshop on modeling of the Arctic atmosphere. *Bulletin of the American Meteorological Society, 86*, 845–852.

Ware, R., Cimini, D., Campos, E., Giuliani, G., Albers, S., Nelson, M., … Cober, S. (2013). Thermodynamic and liquid profiling during the 2010 Winter Olympics. *Atmospheric Research, 132–133*, 278–290.

Westwater, E. (1978). The accuracy of water vapor and cloud liquid determinations by dual-frequency ground-based microwave radiometry. *Radialogic Sciences, 13*, 677–685.

WMO. (2010). *Guide to meteorological instruments and methods of observation*. Technical Note, No. 8. 17 pp.

WMO/CIMO. (1991). *International organizing committee for the WMO solid precipitation measurement intercomparison*. Final Report of the Fifth Session. Geneva: Valdai, Russia. WMO, 31 pp..

Yang, D. (2014). Double fence intercomparison reference (DFIR) vs. bush gauge for "true" snowfall measurement. *Journal of Hydrology, 509*, 94–100.

Yang, D., Benson, C. S., & Ishida, S. (1998). Adjustment of daily precipitation data at 10 climate stations in Alaska: application of World Meteorological Organization intercomparison results. *Water Resources Research, 34*, 241–256.

Yang, D., Goodison, B. E., Metcalfe, J. R., Louie, P., Leavesley, G., Emerson, D., … Milkovic, J. (1999). Quantification of precipitation measurement discontinuity induced by wind shields on national gauges. *Water Resources Research, 35*, 491–508.

Yang, D., Ishida, S., & Gunther, T. (1999). Bias correction of daily precipitation measurements for Greenland. *Journal of Geophysical Research, 104D*, 6171–6181.

Yang, D., & Ohata, T. (2001). A bias corrected Siberian regional precipitation climatology. *Journal of Hydrometeorology, 2*, 122–139.

Zhang, Y., Ohata, T., Yang, D., & Davaa, G. (2004). Bias correction of daily precipitation measurements for Mongolia. *Hydrological Processes, 18*, 2991–3005. http://dx.doi.org/10.1002/hyp.5745. Published online in Wiley InterScience www.interscience.wiley.com.

Zhang, L., Zhao, L., Xie, C., Liu, G., Gao, L., Xiao, Y., … Qiao, Y. (2015). Intercomparison of solid precipitation derived from the weighting rain gauge and optical instruments in the interior Qinghai-Tibetan plateau. *Hindawi Publishing Corporation, Advances in Meteorology, 2015*, 11. http://dx.doi.org/10.1155/2015/936724. Article ID 936724.

INDEX

'*Note*: Page numbers followed by "f" indicate figures and "t" indicate tables.'

Q

R

S

Edwards Brothers Malloy
Ann Arbor MI. USA
November 3, 2016